T0133454

# Programming Language Fundamentals by Example

# Programming Language Fundamentals by Example

D. E. Stevenson

 Auerbach Publications
Taylor & Francis Group
Boca Raton   New York

Auerbach Publications is an imprint of the
Taylor & Francis Group, an informa business

Auerbach Publications
Taylor & Francis Group
6000 Broken Sound Parkway NW, Suite 300
Boca Raton, FL 33487-2742

© 2007 by Taylor & Francis Group, LLC
Auerbach is an imprint of Taylor & Francis Group, an Informa business

No claim to original U.S. Government works
Printed in the United States of America on acid-free paper
10 9 8 7 6 5 4 3 2 1

International Standard Book Number-10: 0-8493-7016-7 (Hardcover)
International Standard Book Number-13: 978-0-8493-7016-8 (Hardcover)

**Library of Congress Cataloging-in-Publication Data**

Stevenson, D. E.
  Programming language fundamentals by example / D.E. Stevenson.
    p. cm.
  Includes bibliographical references and index.
  ISBN 0-8493-7016-7 (alk. paper)
   1. Computer programming--Textbooks. 2. Programming languages (Electronic computers)--Textbooks. I. Title.

  QA76.6.S746 2006
  005.1--dc22                                          2006048426

**Visit the Taylor & Francis Web site at**
**http://www.taylorandfrancis.com**

**and the Auerbach Web site at**
**http://www.auerbach-publications.com**

# CONTENTS

# PART I: MILESTONES

# PART II: GENERAL INFORMATION

# LIST OF FIGURES

# THE AUTHOR

Professor D.E. (Steve) Stevenson is an associate professor of computer science at Clemson University, where he is also the director of Institute for Modeling and Simulation Applications. He received a B.A. in mathematics in 1965 from Eastern Michigan University, an M.S. in computer science from Rutgers University in 1975, and a Ph.D. in mathematical sciences from Clemson in 1983.

He was an infantry officer for four plus years from 1965–1969. He spent one year in Korea and one year in Viet Nam, where he was an advisor to the 3/3 Battalion, 1st ARVN Division. After leaving the military, he was employed at Bell Telephone Laboratories from 1969–1980. While at Bell Labs, he worked on system support, IMS development (loaned to IBM), and as an internal consultant on performance and database design. Dr. Stevenson worked with early versions of C and Unix. He also worked in many interdisciplinary activities involving modeling and simulation; such activities are now known as computational science and engineering.

Computational science led Dr. Stevenson to leave Bell Labs in 1980 to pursue a Ph.D. degree in mathematical sciences at Clemson, specializing in numerical analysis. Since arriving at Clemson, he has worked to develop computational science concepts. Along with Drs. Robert M. Panoff, Holly P. Hirst, and D.D. Warner, he founded the Shodor Education Foundation, Inc. Shodor is recognized as a premier computational science education materials developer and is known for its education of college and university faculty in authentic modeling and simulation activities for science education. The Shodor activities led Dr. Stevenson to investigate new pedagogies; hence, this text is the culmination of some 40 years of experience in interdisciplinary educational and professional activities.

This work is driven by two sayings, the first attributed to Confucius:

"I hear, and I forget.
I see, and I remember.
I do, and I understand."

The second saying comes from Plato: "The unexamined life is not worth living."

# PREFACE

This book is the outgrowth of approximately ten years of experimentation in education. It began when I became dissatisfied with the performance of undergraduate students in computer science studies. It seemed to me that the students were not involved in the process and that their professional skills were not improving. From that point onward, I have found myself far more involved in questions of psychology and educational pedagogy than the material itself.

Although the technical aspects of programming languages have advanced in the past ten years, students still start from the same base of knowledge. I found that students had little understanding of natural language grammars and even less understanding of semantics and pragmatics. Such a lack of understanding meant that the students only had a superficial understanding of language in the broader sense of the word. This lack of understanding of language means that the students have little idea of what language does and therefore how to write a program that transforms language. Therefore, this project starts with an introduction to linguistics.

Before retreating to the hallowed halls of academia, I had the good fortune to work at the "old" Bell Telephone Laboratories. It was my further good fortune to work in a group that had a working relationship with the Unix development folks in Murray Hill, people like Ken Thompson and Dennis Ritchie. This association colored—as it has most of a generation of computer scientists—my perception of language and systems. I have tried to give this text a flavor of the type of decisions that are faced by language designers by answering questions about language internals with "What does C do?" Although I use a broad spectrum of languages in my work, including Fortran 95, Prolog, and OCaml, our students have a very limited experience base. Asking them to consider the details of C gives them practice in using their own powers of deduction.

The second break at Bell Labs was an assignment to IBM to participate in IMS development and a subsequent assignment at Bell Labs in a "consultant" organization. The upshot of this experience is the emphasis in this text on project management. Judging by the numbers of computer science undergraduates versus graduate students, it is clear that most undergraduates must find work in industry. In industry, the rules of the game are much different than those in academia. Anecdotal evidence with students shows that they have very poor time management skills and even fewer design skills. Although most students have at least one software engineering course, it is rare to find software engineers interested in design. Design is learned by designing.

This is a long way back to the opening idea: students were not engaged and I needed to find a new way of teaching to get the real-world flavor. That new approach is really quite old: problem-based learning and case-based methods. These are two of a group of methods known as inquiry-based or active learning methods. Although I hope this text is a shining example of how to use these ideas in computer science education, I hasten to point out that there is much to be learned about computer science education pedagogy. I urge you to begin by reading *How People Learn* (Bransford, Brown, and Cocking 2000). It is more than worth every second to read it.

## AVAILABILITY OF COURSE MATERIALS

1. **Gforth Installation.** This text uses Gforth as the target of the project. Actually, any Forth will do but Gforth is widely available—for free—and is under maintenance. If you are not familiar with the GNU project, it is recommended that you visit their Web site at www.gnu.org. Instructions on downloading Gforth are found on the Web site.
2. **Forth Documentation.** There are two major documents supporting Gforth: the 1993 American National Standards Institute (ANSI) Standard and the *Gforth Reference Manual.* Both documents are available at www.cs.clemson.edu/steve/CRCBook.
3. **PSDP Forms.** The Microsoft® Excel spreadsheets used for the project management portion of the project are available at www.cs.clemson.edu/steve/CRCBook.

# 1

## A WORD ABOUT USING THIS TEXT

This text, I believe, is unique in the computer science education market. Its format has been carefully thought out over approximately five years of research into problem-based and case-based learning methodology. This is a student-centered text: there is no attempt to make this a typical university textbook. There are many very good textbooks on the market that focus on encyclopedic inclusion of issues in the development of programming languages; this is *not* one of them.

*Student-centric* can mean different things to different people. This text is meant to be read by students, not faculty. This distinction means that the text presents problems to be solved and then presents the information needed by the student to solve the problem. The tone is informal and not academic. The ultimate purpose of the text is to show students how to be successful computer scientists professionally. At Clemson, this is considered our capstone course.

## 1.1 EXPECTATIONS FOR THE STUDENT AND INSTRUCTOR

To start the text I will give my assumptions and goals for the project.

1. Students are competent in one program from the C class of languages (C, C++, or Java). Their competence is assumed minimal at the professional level because they have not spent enough time to gain expertise.
2. Students are familiar with Unix.
3. Students are comfortable using search engines. A running joke in my classes is "Google is your friend."

If students will faithfully follow the project in this text, then experience teaching this course at Clemson shows the following:

1. The students will develop a full-scale project using professional-grade development management techniques. Anything less than full participation and the students will not gain all this text has to offer.
2. The students will master concepts, languages, and experiences to prepare them to understand the current research in programming languages.

In keeping with the problem-based nature of the text, here is the one problem you want to solve:

> What theory and knowledge do you need to develop a compiler for a higher-level language? What personal and professional methods do you need to successfully develop a large project? You're successful if you can explain all of the steps and algorithms to compile the following program and successfully run it:

```
[define :[g :[x int] int]
        [if [= x 0] 1 [* x [g [- x 1]]]]]
```

## 1.2   OPENING COMMENTS

This text is student-centric because it does not try to be encyclopedic, but rather guides students through the development of a small compiler as a way for them to understand the issues and commonly used approaches to the solutions of those issues. For this reason, the text is written in an informal, chatty style. I am attempting to set up a professional, design-team atmosphere and to promote strategies that will work in professional practice.

One issue that I have noticed uniformly across years, disciplines, and students is a complaint that textbooks are boring and therefore students do not read as they should. On closer examination, what one finds is that students rarely master fundamental study skills. In my classes I encourage by the way I make assignments what I called the FSQRRR method: Focus issue, Survey, Question, Read, Recite, Review. SQRRR is a familiar technique in K–12 education but not at the university level. The student first needs to understand the issues. Students often do not develop a questioning attitude and, therefore, when they do read, there is no focus or purpose.

The planning of the reading is a metacognitive exercise. There are many good online sites that describe SQRRR*; I encourage students and teachers alike to survey this technique and apply it to course work.

Without going into too much detail, a word about problem-based learning. There is a wealth of research that students learn best when they have a realistic problem before them and that problem *must* be solved. Please read the National Research Council's *How People Learn: Brain, Mind, Experience, and School* (2000). The research indicates that the problem should be ill-structured. Certainly, asking students to write a compiler for a language they have never seen qualifies as "ill-structured" and I claim it is a "realistic" expectation that a senior/post-graduate be able to design and implement such a compiler *if* the proper guidance is given. Therefore, this text seeks to provide guidance.

I have made a distinction in the planning of this text not usually drawn in the education literature. Problem-based learning is generally taken as "small" problems, hardly the adjective for a compiler project. The ideas of problem-based learning come under a more general term, *inquiry-based learning*. There are several other forms of inquiry-based learning but the rubric of *case-based learning* has a connotation separate from problem-based learning. Case-based learning is used in business and the law and stresses the judgmental issues of problem solving. Computer science is filled with judgments throughout every design. Every problem has many ways to be solved, each with its own computational costs in time and space. Computational costs are often not known in detail but require judgments by the designer. I have planted many case-based issues throughout the project: treat these as a minimal set; any judgment call can be discussed in a case-based rubric.

It is customary in case-based situations to provide commentary on the case. Such commentaries, if the students were to get them, would subvert the entire learning process. For that reason, the *Instructors' Manual* is a separate book, keyed to the cases in the student edition.

The approach of this text stimulates broad participation in the computer science enterprise because it emphasizes self-education and self-evaluation, as well as the design and implementation of complex systems. Just as metacognition is an important lifelong learning technique, so is explanation. I encourage instructors and students alike to spend time explaining the details of the problem at hand to one another. I like to emphasize an observation credited to Albert Einstein: "You don't understand something unless you can explain it to your grandmother." This observation emphasizes the idea that difficult technical ideas must be

---

* Google for it.

explainable in simple, ordinary English (or whatever natural language you're using).

The last educational issue that I want to discuss is the question, "How do I know the learner has learned?" My view is that in computer science, you only know something if you can present a working algorithm that produces only correct answers. I encourage the users of this text to spend time doing metacognitive exercises. Metacognition plays a crucial role in successful learning. Metacognition refers to higher-order thinking, which involves active control over the cognitive processes engaged in learning. Activities such as planning how to approach a given learning task, monitoring comprehension, and evaluating progress toward the completion of a task are metacognitive in nature. As Plato said, "The unexamined life is not worth living."

Along with the FSQRRR issue there are several issues concerning design of large software systems. There are at least two different aspects that should be addressed:

1. In light of the comment about metacognition, students should be urged to take stock of their practices. This is not a text on the personal software process (PSP) outlined by Humphrey (1996), but I have included a chapter on PSP, augmented with some design forms that I use in my class. My students—and, I shamefacedly admit, I—do not make good use of our time, and mostly because we do not take stock of how our time is spent. Do yourselves a favor: get a firm grip on your time—it's your only capital.

2. Current computer science curricula tend to focus on single procedures and not software systems. Therefore, students' design skills are fairly rudimentary. The project inherent in this text is far too complicated to give without much guidance. There is a fine line between "doing all the intellectual work for them" and "allowing them to figure things out for themselves." The goal is that the students learn how to do decomposition, analysis, and synthesis but my experience is that we, the instructors, need to give them guidance. I often quote George Pólya (1957): "If you cannot solve this problem, there is a simpler problem in here you can solve. Find and solve it." But I warn you, the students don't find this helpful the first time(s) they try to use it.

## 1.3 POSSIBLE SEMESTER COURSE

This section provides guidance in using this text and one possible sequence of material (see Figures 1.1 and 1.2).

| | *Class Periods* | | |
|---|---|---|---|
| *Week* | *Period 1* | *Period 2* | *Period 3* |
| 1 | Opening. Get data sheet. Explain how and why this course is different. Assign the specification as reading. | Organize groups. Organize Q&A on specs. Assign in-class exercise to code Hello World. Start *one-minute write* habit. Assign Chapter 5. Set Milstone I due date. | Address the focus question. Use the discussion to develop study plan. Assign Chapter 6 to focus question "How are programming languages like natural language?" |
| 2 | Explore the focus question. Assign Chapter 7. | This and the next period should pull together a complete block diagram of a compiler. Assign Chapter 8. | Ibid. |
| 3 | Work through the calculator. Assign Chapter 9. | Ibid. | Wrap up. Discuss overview of *Gforth*. Case 2 Introduction for Milestone I |
| 4 | Case 3. | Case 4 and 6. | Milestone I due. |

**Figure 1.1   Sample Organization for Introduction**

## 1.3.1   General Planning

Good educational practice outlined in *How People Learn* (Bransford, Brown, and Cocking 2000) dictates that the course be built around three learning principles: (1) prior knowledge—learners construct new knowledge based on what they already know (or do not know); (2) deep foundational knowledge—learners need a deep knowledge base and conceptual frameworks; (3) metacognition—learners must identify learning goals and monitor their progress toward them.

### 1.3.1.1   *Learning Principle 1: Prior Knowledge*

The course I teach primarily requires deep knowledge of data structures. This deep knowledge includes not just the usual fare of lists, trees, and

| | Class Periods | | |
| --- | --- | --- | --- |
| Week | Period 1 | Period 2 | Period 3 |
| 5 | | | |
| 6 | | | |
| 7 | | | Milestone II and III due |
| 8 | | | |
| 9 | | | Milestone IV due |
| 10 | | | |
| 11 | | | Milestone V due (passing grade) |
| 12 | | | |
| 13 | | | Milestone VI due |
| 14 | | | |
| 15 | | | Milestone VII: entire project due |
| 16 | | Metacognitive final | |

**Figure 1.2  Sample Milestone Schedule**

graphs, but also the ability to

1. Formulate detailed time and space requirements
2. Implement fundamental-type systems such as strings
3. Formulate alternative designs and choose an optimal design based on detailed considerations

I find my students have only surface knowledge of data structures. It would be helpful if the students have had a computability course that emphasizes formal languages. We have such a course at Clemson and about half my students have completed that course. I find that such courses may not be sufficient unless it, too, is a problem-based learning course.

Learners must have some architectural experience. I find that assembler-level programming, even among computer engineering students, is on the wane. My use of Gforth in my course was prompted by *Computer Architecture and Organization: An Integrated Approach* (Heuring and Jordan 2003) because the hardware description language in the text is easily simulated in Gforth.* Expert programmers know that the implementation language is just a language. Seniors (and graduate students) in computer science are unlikely to be that experienced, so mastering Forth represents a true departure from the steady C diet.

Without trying to compete with Watts Humphrey's views expressed in the *Personal Software Process* (1996), I find that students have no software

---

* A newer (2006) edition of this title is available but I have no experience with the text.

process—period. In particular, students have no well-developed idea of testing. This spills over into learning principle 2.

In summary, although the students all believe they know these subjects, the majority will not have deep operational experience. You must leave room in the syllabus for in-depth design and implementation. A perennial problem in my classes is Gforth support of the primitive `string` type.

### 1.3.1.2 Learning Principle 2: Foundational Knowledge

In discussions of knowledge, the type of knowledge referred to here is *implicit* knowledge. Implicit knowledge is held by the individual and by definition is not something that can be written down. In the expertise literature, implicit knowledge is what we ascribe to experts that make them experts.

Here, too, I find the students struggle. The expertise literature suggests that it takes about 10,000 hours of effort to gain expertise in a subject; that's about five years. Things that are obvious to the instructor are generally not obvious to the students due to this novice–expert difference. For example, an experienced instructor can reel off several alternative designs; novices struggle to get one.

This implicit knowledge comes into play at the higher levels of work: design. Figure 1.3 shows three considerations that are important in the education literature. Bloom's taxonomy (column one) is regarded by education specialists as a seminal insight; despite its age, the basic breakdown is useful. Instructors tend to work at the evaluation–synthesis–analysis level, whereas students tend to work at the application–comprehension–knowledge level. Anderson and Krathwohl (2001) reworked the taxonomy (columns two and three): (1) conceptual and factual knowledge are learning principle 1; (2) implicit knowledge is learning principle 2; and (3) metacognitive knowledge is learning principle 3. Implicit knowledge

| | | Metacognitive Knowledge |
|---|---|---|
| Evaluation | Create | |
| Synthesis | Evaluate | Implicit Knowledge |
| Analysis | Analyze | |
| Application | Apply | |
| Comprehension | Understand | Conceptual Knowledge |
| Knowledge | Remember | Factual Knowledge |

**Figure 1.3  Bloom's Taxonomy of Educational Objectives (Data from Bloom 1956; Anderson and Krathwohl 2001.)**

(also called procedural knowledge) is the knowledge of *how* to do things: play the violin, add two columns of numbers, and recognize patterns.

### 1.3.1.3 Learning Principle 3: Metacognition

Metacognition is informally defined as "thinking about thinking." George Pólya (1957), for example, in the classic *How to Solve It*, emphasized that thinking about the solution of a problem was the last step. I use metacognition in two basic ways in this course. Every class period each student fills out a *one minute write* (OMW) exercise using a $3 \times 5$ scratch pad. Students answer two questions: (1) "What did you learn today?" and (2) "What questions do you have?" OMW exercises count as a participation grade *and* serve as a peek into the minds of the students. I consider this as a privileged communication with the student. I read them all and comment on them. The questions can get far off track, but they can also help the instructor spot troubled students.

The second way I use metacognition is in milestone reports. The milestone reports in general are factual, such as how the student exactly used a particular technique. It also contains a metacognitive exercise, "What did you learn?" I always look forward to this section, because—at least in my experience—the students are usually totally honest. The milestone report is 25 percent of the milestone grade so the students cannot take it lightly. I have included the reports of one of my A students to indicate what I consider a good series of reports (see Appendix A).

## 1.4 DETAILED SEMESTER PLAN

### 1.4.1 Milestone Maps

The organizational feature of the course is the milestone. The milestones interact as shown in Figure 1.4. Note that Milestone I is actually the last one in line as far as the total project is concerned, but it is done first so that the students understand what they are compiling to. In terms of more compiler-conventional terminology, the milestone control and data flow is shown in Figure 1.5.

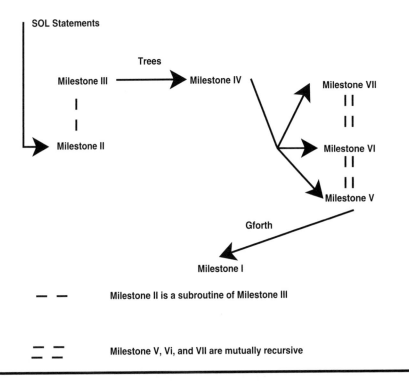

**Figure 1.4    Relationship of the Milestones**

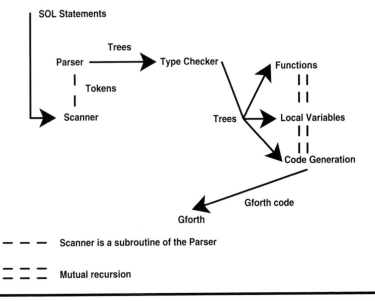

**Figure 1.5    Control and Data Flow**

# 2

## INTRODUCTION TO THE PROBLEM

### 2.1 MEMORANDUM FROM THE PRESIDENT

Congratulations on being named to the company's secret compiler project. Alternative Arithmetics is pleased to have developed a stack-based processor that achieves five petaflops. Such processor speeds make many heretofore impossible computations possible.

In order to make use of this speed, we must develop a new compiler. We have considered the three types of languages: imperative, functional, and logical. We believe our best chance for success lies in developing a functional language that we are tentatively calling *Errett* after the mathematician Errett Bishop. Functional languages generally allow for very powerful type systems (*ML*, for example) and very clean semantics.

In order to proceed on parallel development paths, AA's research has developed a simple version of the language design that can produce information in the appropriate form for you. We have decided to take the same tack as the GNU gcc compilers by separating the syntactic processing from the semantics and code generation segments. Your group will be developing the second phase: the semantics processing and code generation aspects. There are not enough chips available for you to work with a live chip, but the new petaflop chip uses a Forth-like input language. Therefore, you will be converting the syntax trees developed for Errett into Forth code. Because of its stable nature, we have decided to use the GNU Gforth system as the emulator. We have also provided a prototype syntactic processor so you can experiment with the interface.

Other memos are in preparation that outline the specific trees that you will have to process and the semantics of each of the operators or constructs.

Best of luck. We're happy to have you here at Alternative Arithmetics and we're counting on you to develop the prototype compiler necessary to our success.

Best regards,
Socrates Polyglot
President

## 2.2  BACKGROUND

The Simple Object Language (*SOL*), pronounced "soul," is the primitive language for the new Simple Object Action Processor (*SOAP*). The SOAP processor is under development but we can have software emulation by using the Gforth system. Some features of Gforth may not be available in the SOAP.

### 2.2.1  Cambridge Polish

The prototype you are working on has two top-level constructs: type definitions and `let` statements. The output from these two constructs are linearized versions of their parse trees. Such a format has been long used in the Lisp language, dating back to the mid-1960s, and this format has long been called "Cambridge Polish."

Cambridge Polish is quite simple to understand and its beauty is that it is an efficient way to encode the structure of trees. As you all know, programs can be represented by trees. Let's recall some properties of trees:

- A *binary tree* is a graph, $T = (N, E)$, where $N$ is a set of *nodes* and $E$ is a subset of $N \times N$. Nodes can be anything and recall that the expression $N \times N$ is shorthand for "all possible pairs of nodes." Technically, $E \subseteq N \times N$. An *n-array* tree would be a generalization of a binary tree.
- A tree can be *directed* or *undirected*. From your data structures class you may recall that directed graphs can only be traversed in one direction. Cambridge Polish trees are directed.
- Trees are connected (no isolated nodes) and have no simple cycles. That terminology should be familiar to you from your data structures class.
- Trees can be *rooted* or *unrooted*. Cambridge Polish trees are rooted.
- Any two vertices in $T$ can be connected by a unique simple path.
- Edges of rooted trees have a natural orientation, toward or away from the root. Rooted trees often have an additional structure, such as ordering of the neighbors at each vertex.

■ A labeled tree is a tree in which each vertex is given a unique label. The vertices of a labeled tree on $n$ vertices are typically given the labels 1, 2, ..., $n$.

The syntax for Cambridge Polish is simple. There are three types of symbols: the left bracket '[', the right bracket ']', and atoms. Atoms can be one of many different types of strings—integers or strings or real numbers or variable names—but to the syntax they are all the same. The left bracket indicates the beginning of a node and the right bracket indicates the end of a node.

Examples include:

[] is the node with nothing in it—NIL in some languages.
[1] is a node with just one atom, the integer 1.
[+ 1 2] is a node with three atoms: '+', 1, and 2.
[+ [* 2 3] 4] is two nodes (because there are two left brackets). There are five total atoms: '+', '*', 2, 3, and 4. This represents a tree that could be drawn as shown in Figure 2.1.

## 2.2.2 A Note on Notation

When describing the syntax below, we use two forms of notation. Atoms that appear like-this are written exactly as shown. Atoms that appear *like-this* can be thought of as variables. For example,

[let *variable value*]

should be read as follows: The [ and the word let are written as shown; *variable* can be replaced by any allowed variable and *value* can be replaced by any valid expressions. The last statement represents the tree shown in Figure 2.2.

In Cambridge Polish the first element is considered to be the function name and the other elements are arguments.

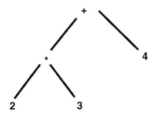

Figure 2.1   Tree for [+ [* 2 3 ] 4 ]

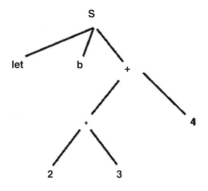

**Figure 2.2 Drawing of the Tree [let b [+ [* 2 3] 4]]**

### 2.2.3 A Note on Names

There are several uses for what we normally call *variables* or *names* in programming. Many key words and reserved words in programming are "spelled" just like variables. These are *surface spellings*, meaning they are written and input as shown. Internally, though, there are many names that we want to use but must be sure that these names are not something that can be confused with names in the program. We call those *pseudo-functions* and these names are preceded by the number sign (Unicode 0023) ('#'). There are design issues to be decided upon in constructs using these pseudo-functions. For example, an array constant is typed by

$$[1,2]$$

but it is presented to your input as

$$[ \text{\#arraydef } 1 \ 2 \ ]$$

## 2.3  *SOL*

SOL is a cross between an assembler language and a higher-level language. Languges such as SOL were in vogue in the late 1960s and early 1970s (PL360, for example). It adopts a simple syntactic structure, but it also adopts simple statements such as expressions and assignments and the use of higher-level control statements such as if and while.

The current state of the SOL syntactic processing is described here. The output of the syntactic program is in Cambridge Polish, and the semantic intent is described in this memo. All statements begin with an atom that designates the function to be applied. So, for example,

$$[+ \ 1 \ [* \ 2 \ 3]]$$

is the Cambridge Polish rendition of the tree for $1 + 2 * 3$. Some functions such as `let` are pseudo-functions because they take their meaning at compile time, not run time.

Generally, labels, function names, and type names can be any valid C variable name but without the underscore ('_'). Primitive operations use conventional symbols. Operands can be variable names, C integers, C floating point numbers, or C strings. Operands can also be array and pointer expressions. The variable names `true` and `false` are reserved for Boolean values and hence cannot be used as function or variable names.

The language is primarily functional in orientation: every primitive operator returns a value except the function definition; and `while`. `if` statements are a bit different than the standard C one: `if` can return a value and therefore when used functionally, both branches of the `if` must return the same value. See, too, the comment above on the use of names that begin with the number sign.

Comments do not appear in the Cambridge Polish version.

## 2.4 PRIMITIVE DATA AND PRIMITIVE OPERATIONS

The elements of the SOL are much like C: booleans, integers, floating point numbers, strings, and files. The language also allows for variables and function symbols.

### 2.4.1 Constants

**bool:** Boolean: true and false
**int:** Integers: C-type integers
**float:** Floating Point: C-type double
**string:** Strings: C-type strings
**file:** Files

The words in bold code font are the names of primitive types.

### 2.4.2 Variables

Variables and functions are defined by `let` statements. There are two types of `let`s. One is a "top level" that is roughly equivalent to defining globals in C. The format for the top level `let` is

[let [*definiendum definiendum*]... ]

The words in italics are called *meta-variables* and those in bold are keywords; keywords must be written exactly as shown. *Definiendum* is

Latin for "the thing being defined" and *definiens* is Latin for "the defining thing." The *definiendum* can be one of two forms: a variable or a function.

In order to define a simple scalar variable, we use the following format:

$$variablename : type$$

where *variablename* is a new name otherwise undefined and *type* is one of the allowed types: bool for Boolean, int for integers, float for doubles IEEE 754 constants, and string for strings.

### 2.4.2.1 Arrays

Array notation similar to C is used in the language. This translates into terms: the bracketed notation [3,5] is translated into [#arrayapp 3 5]. On the other hand, if we define an array, we see something quite different:

$$[\#arraytype \ type \ shape \ ]$$

where *type* is the type of the array and *shape* is the number of rows, columns, etc.

### 2.4.2.2 Assignment

We will have an assignment statement

$$[ := lvalue \ rvalue \ ]$$

The lvalue can be an expression that evaluates to a pointer ("lvalue") and the rvalue is any expression ("rvalue") that evaluates to data. The lvalue and rvalue must be type compatible. In the early stages (before Milestone VIII), we can have a simple approach to assignments and the evaluation of variables. In the complex data types, we will revisit the simple evaluation.

## 2.5 PRIMITIVE DATA OPERATIONS

Each of the primitive data types comes equipped with pre-defined operations. The tables shown in Figures 2.5 to 2.12 later in the chapter list the primitive operations.

### 2.5.1 Stack Control Operations

Because the chip uses the Forth paradigm, certain of the stack control operations are available to the programming—but not all.

### 2.5.1.1 Stack Position Variable

Each Forth stack can be addressed by position. The general format of a stack position expression is a dollar sign ($) followed optionally by one letter followed by a number. The letter can be A (argument) or F (floating); if no stack is specified, then A is assumed. The numbers can be any non-negative integer, with 0 being the top of stack (TOS). If the number is missing, then the TOS (position 0) is assumed. Examples are

- $ is the top of the argument stack. $F$ is the top of the floating point stack.
- $5 is the *sixth* item in the argument stack.
- $F6 is the *seventh* item in the float stack.

### 2.5.1.2 Stack Operations

The SOAP processor is a stack-based machine with a number of standard stack operations that are described in the ANSI Forth manual: drop, dup, fetch, over, pick, roll, store. Note that stack operations may be implied by the stack position operator.

## 2.6 CONTROL STRUCTURES

There are three basic control structures: sequential, selection, and repetition.

### 2.6.1 If-Selection

The if control structures have the general forms of

[if *condition statement1 statement2* ]

[if *condition statement1* ]

These are exactly like the C standard except that the *condition* must have a Boolean type.

### 2.6.2 Select-Selection

The select statement is a generalized *if* statement. The sequential mode does not work here: not all instructions are to be executed. Therefore, we need markers to indicate the end of the "true" side and the end of the "else" side. Nested *if* statements can be quite messy, so we add a purely human consideration: how can we tell which *if* section ends where? The `select` statement is similar to the `switch` in C, with the exception that the code executed when the condition is true is treated as if it ended with `break`. *select* also differs from C in that the case conditions are arbitrary expressions, not just constants.

[`select` *expression caselist* ]

where *expression* computes a value with the usual interpretation. The *caselist* is a list of pairs:

[`#selectcase` *test result*].

Each *test* is computed in the order stated. If the *test* returns `true`, then the *result* is computed. If the *test* is `false`, then the next pair is evaluated. This is similar to C's *switch* statement, but the *test* can be a general statement.

One of the pairs can be the `default` condition, which should be the last pair.

### 2.6.2.1 While

While statements are syntactically very simple:

[`while` *condition body* ]

This statement's semantics is completely similar to C's *while* statement. The *body* may have zero or more statements.

### 2.6.2.2 Compound Statement: Begin-End

The compound statement can appear anywhere that a simple expression can appear. The need for compound statements is rare in SOL because the *while* statement *body* acts as a compound statement. *if* statements rarely need to use compound statements because those statements must return values.

## 2.7 EXPRESSIONS

SOL is a functional language. This means that virtually all constructs return a value, the major exception being the *while*. Expressions are written in prefix notation and this is the major reason Cambridge Polish is used as the intermediate notation. In general, then, expressions are written as

[ *operator operand1 ... operandn* ]

The list of operations and their attributes is provided later in this chapter starting with Figure 2.5.

One major difference in SOL from C is the fact that *if* and *select* statements must return consistent values because the body of a function might be only a *select* statement.

With regard to stacks, because SOL will work with a stack-based environment, special notation is invented to address the two stacks: the argument-integer stack and the floating point stack. These are represented by $integer and $Finteger, respectively. Reading a stack location does not alter its value; *assigning* to a stack location does alter the value.

Because SOL is a numerically oriented language, arrays are first-class citizens. Array constants are represented as

[#arrayapp *element1 ... elementn* ]

Functions are defined using let statements. The form of a definition requires some thinking about the Cambridge Polish syntax. In Cambridge Polish, functions are written in *prefix Polish:* the function symbol comes first, followed by the arguments. For example, *a*(*b, c*) would be written [a b c]. Therefore, a function definition would look like

[let [ [a b c] *definiens* ]]

Local variables in functions (in fact, anywhere) are a special form of the let function.

[let   *definiendum definiens expression* ]

Notice that this version of the let has three elements exactly; the functional definition has only two.

## 2.8   INPUT FILE STRUCTURES AND FILE SCOPE

A file may contain multiple function definitions. A file may also contain variable statements. Variable statements outside a function are considered to have global scope outside the file but local scope in the file: a variable is not known until it is defined and it is known from that point to the end of the file.

Files may have executable statements outside function or module definitions. These are taken to be statements to be executed at the time the file is loaded by Gforth.

| | | |
|---|---|---|
| typedefine | → | [ *type* [ #typedef *name typeterm* ] ] |
| typeterm | → | [ #bar *type typeterm* \| *type* ] |
| type | → | primitive \| |
| | | [ #of *name typeterm* ] \| |
| | | *name* \| |
| | | *typevariable* \| |
| | | [ #tupletype *typelist* ] \| |
| | | [ #ptrdef *type* ] \| |
| | | [ #arraydef *commalist* ] |
| | | [ #fundef *type type* ] \| |
| | | [ #object *complexinput* ] \| |
| | | [ #struct *labeldtypelist* ] |

**Figure 2.3   Type Sublanguage**

## 2.9   DERIVED TYPES

SOL provides for derived types by the type statement typename: type.

$$[\text{type } \textit{type-expression } ]$$

The *type-expression* is a sublanguage in its own right (Figure 2.3).

Classes are derived types with inheritance, information hiding, and polymorphism. Objects are instances of classes (Figure 2.4). Various conventions based on Java are assumed. As usual, constructors take on the class name and are polymorphic.

Dynamic allocation is provided using allocate and deallocate, precisely like "malloc" and "free" in C. A pointer value is dereferenced using the "dot" notation outlined above. *allocate* has one argument, a type

```
[class  →   class-name
            [import name-list]
            [public name-list]
            [private name-list]
            optional type declaration
            optional let statements for variables
            . . .
            methods
            optional let statements for variables
            . . . ]
```

**Figure 2.4   Class Syntax**

name from which the space needs are created. `allocate` returns a valid pointer if space is available and 0 if space is not available. The built-in function `isnull` returns true if a pointer is 0 and false if the value is a non-zero pointer.

## 2.10   SCOPE AND PERSISTENCE

There are two scope types in SOL: file scope and function/compound statement scope. In general, definitions in either scope are considered known from the end of the definition statement until the end of the file (in file scope), function (in function scope), or immediately after the `begin` (in compound statement scope).

Variables defined in file scope are considered static and persist throughout the lifetime of the program. Variables defined in functions local in scope and transient (on the stack). Function names are always considered static.

### 2.10.1   Scope Definition

All declarations outside of functions are considered to be in the file scope. File scope is defined as follows:

- Every definition in file scope is considered local to the file unless named in an export statement. Variables, functions, and types may be shared in this way.
- Definitions from other files may be used; however, declarations must appear in an export statement in the other files. A declaration can only occur in one export statement. The using (non-global) file must use an import statement.

An export statement has the form

[export *name1 name2* ]

Each name must be defined in the file in which the export statement occurs.
An import statement has the form

[import *filename name1 name2* ]

The filename must be a valid SOL file and each name must be mentioned in an export statement.

### 2.10.2   Scoping within Functions and Compound Statements

Scoping in functions and compound statements is lexical. This means that there can be nesting of scoping (functions within functions and compound statements within functions within compound statements, *ad nauseam*). Such nesting leads to the "nearest definition" use, as in C (Figure 2.5).

| Operator Name | Operator Symbol | Semantics |
|---|---|---|
| | *Primitive Data* | |
| | *When no Semantics are given, C is intended* | |
| | *Boolean* | |
| And | $A \& B$ | Same as && in $C$ |
| Or | $A \mid B$ | |
| Not | $!A$ | |
| Implies | $A => B$ | False when $A$ is true and $B$ is false |
| Biconditional | $A <=> B$ Logical equals | |

**Figure 2.5    Booleans**

## 2.11    TYPE SYSTEM

The intent is that SOL will strongly become a type language. Unfortunately, development schedules for the early versions of the parser will not allow us to produce the code necessary to do type inference. Therefore, we will fall back on the C-type languages and require every user name to be defined using primitive or user-defined types. The type system should act like C.

## 2.12    PRIMITIVE FUNCTIONS AND ARGUMENTS

The normal precedence conventions hold as shown in the tables below (Figure 2.6 and Figure 2.7).

| | *Integers* | |
|---|---|---|
| Plus | $A + B$ | |
| Minus | $A - B$ | |
| Unary Minus | $-A$ | |
| Times | $A * B$ | |
| Divide | $A/B$ | |
| Remainder | $A \% B$ | |
| Power | $A^\wedge B$ | Integer A raised to an integer power B |
| Less | $A <$ | |
| Equal | $A = B$ | Same as C's == |
| Greater | $A > B$ | |
| Less than or equal to | $A \le B$ | |
| Greater than or equal to | $A \ge B$ | |
| Not equal | $A! = B$ | |

**Figure 2.6    Integers**

| Category | Operators | |
|---|---|---|
| Assignment | := | |
| Logical operations | < = != <= = -> >= | |
| Addition | + - \|\| | |
| Times | * / mod | |
| Power | ** (this is right associative) | |
| Negation | (unary) - | |
| Conversions | type-type | |
| | *Floating Point* | |
| Plus | $A + B$ | |
| Minus | $A - B$ | |
| Unary Minus | $-A$ | |
| Times | $A * B$ | |
| Divide | $A/B$ | |
| Remainder | $A\% B$ | |
| Power | $A^{\wedge}B$ | A float, B either float or integer |
| Less | $A <$ | |
| Equal | $A = B$ | Same as C's == |
| Greater | $A > B$ | |
| Less than or equal to | $A \le B$ | |
| Greater than or equal to | $A \ge B$ | |
| Not equal | $A ! = B$ | |
| Sine | $\sin(A)$ | |
| Cosine | $\cos(A)$ | |
| Tangent | $\tan(A)$ | |
| Exponential | $\exp(A)$ | $e^A$ |

**Figure 2.7   Floats**

## 2.13   BUILT-IN VALUES, FUNCTIONS, AND PSEUDO-FUNCTIONS

```
sin cos tan open close endoffile readbool
readint readfloat readstring writebool
writeint writefloat writestring allocate free
deallocate
```

### 2.13.1   Reads, Writes, and File Operations

There are three standard files: stdin, stdout, and stderr. Following Unix convention, these three are always considered opened when the program begins. Attempts to open or close any of the three are ignored.

1. open. open has one argument, a string. This string designates the file to be opened and is interpreted as a Unix path. open returns one or two elements on the stack.

   a. If the top of the argument stack is true, then the next element on the argument stack is the file.
   b. If the top of the argument is false, the file failed to open; there is nothing else placed on the stack by open.

2. close. close closes the file, its single argument, and leaves nothing on the stack.

3. endoffile. endoffile takes a file as its argument and leaves as a Boolean value on the stack: true if the file is at its end; false otherwise.

4. Read. The four read functions (readint, for example) have a file name as their argument; the result is left on the appropriate stack.

5. Write. The four write statements take a file as their first argument and a value of the appropriate type as the second argument and returns the status (true if the write is successful; false otherwise) on the stack.

|  | *Strings* | |
|---|---|---|
| Concatenation<br>strcat in C | $A + B$ | |
| Insert<br>Insert character $B$ at location $C$<br>in string $A$ | insert($A$, $B$, $C$) | |
| Character of | charof($A$, $B$) | Return the $B$th character of string $A$ |
| Less | $A <$ | |
| Equal | $A = B$ | Same as C's == |
| Greater | $A > B$ | |
| Less than or equal to | $A \leq B$ | |
| Greater than or equal to | $A \geq B$ | |
| Not equal | $A! = B$ | |

**Figure 2.8  Strings**

| | *Files* | |
|---|---|---|
| Open | open(*A*) | *A* is a path expression as a string. Returns a file value |
| Close | close(*A*) | Closes open file *A* |
| Test end of file | endoffile(*A*) | Returns true is file *A* at end of file |
| Read boolean value | readbool(*A*) | File*A* is read for boolean |
| Write a boolean value | writebool(*A, B*) | Write boolean value *B* to file *A* |
| Read integer value | readint(*A*) | File*A* is read for integer |
| Write a integer value | writebool(*A, B*) | Write integer value *B* to file *A* |
| Read float value | readfloat(*A*) | File*A* is read for float |
| Write a float value | writefloat(*A, B*) | Write float value *B* to file *A* |
| Read string value | readstring(*A*) | File*A* is read for string |
| Write a string value | writestring(*A, B*) | Write string value *B* to file *A* |
| | *Special Values* | |
| Pi | Pi | Closest representable value of $\pi = 3.14\ldots$ |
| Standard in | stdin | |
| Standard out | stdout | |

**Figure 2.9   Files and Special Constants**

| | | |
|---|---|---|
| *validstatement* | → | *toplet* \| *typedefine* |
| *toplet* | → | [LET [*oper expression*]] |
| *typedefine* | → | [*type* [#typedef *name typeterm*]] |
| *typeterm* | → | [#bar *type typeterm* \| *type* ] |
| *type* | → | primitive \| |
| | | [#of *name typeterm*] \| |
| | | *name* \| |
| | | *typevariable* \| |
| | | [ #tupletype *typelist* ] \| |
| | | [ #ptrdef *type* ] \| |
| | | [ #arraydef *commalist* ] \| |
| | | [ #fundef *type type* ] \| |
| | | [ #object *complexinput* ] \| |
| | | [ #struct *labeldtypelist* ] |
| *labeledtypelist* | → | *labelentry* \| *labelentry* |
| *labelentry* | → | [*name typeterm*] |
| *complexinput* | → | *validstatement complexinput* |
| *expression* | → | *oper* \| *statements* |

**Figure 2.10   Input Grammar, Part 1**

| oper | → | [ := *oper oper* ] \| |
|------|---|----------------------|
| | | [ : *oper type* ] \| |
| | | [ *binaryops oper oper* ] \| |
| | | [ *unaryops oper* ] \| |
| | | [ *oper oper* ] \| |
| | | *constants* \| |
| | | [ LAMBDA *lambdanames expression* ] \| |
| | | [#arrayapp] \| |
| | | [ #arrayapp *commalist* ] \| |
| | | [ #tupleapp \| |
| | | [ #tupleapp *commalist* ] \| |
| | | NAME \| |
| | | *STACK* |
| statements | → | *if* \| *while* \| *begin* \| *let* \| *select* |
| if | → | [ IF *expression expression expression* ] |
| | | \| [if *expression expression* ] |
| while | → | [ WHILE *expression exprlist* ] |
| begin | → | [ BEGIN *exprlist* ] |
| let | → | [ LET *expression expression expression* ] |
| select | → | [ SELECT *selectlist* ] |
| selectlist | → | *selectterm* \| *selectterm selectlist* |
| selectterm | → | [ #seleccase *expression expression* ] |
| | | \| [ #selectcase *DEFAULT expression* ] |

**Figure 2.11    Input Grammar, Part 2**

| import | → | [ #import *namelist* ] |
|--------|---|------------------------|
| namelist | → | NAME \| NAME *namelist* |
| export | → | [ #export *namelist* ] |
| commalist | → | *expression commalist* \| *expression* |
| exprlist | → | *expression exprlist* \| *expression* |
| lambdanames | → | [ COLON NAME *type* ] *lambdanames* \| |
| | → | [ COLON NAME *type* ] |
| typelist | → | *typeterm* \| *typeterm typelist* |

**Figure 2.12    *SOL* Input Grammar**

## 2.14    TYPE CONVERSIONS

Type conversions are essentially a one argument function that produces one output. The input is the value to be converted and the output is the value in the new system. Because we would like to have a general ability to do conversions, we adopt a class of functions of the form name1→name2. Whenever this format appears as the operation, the intent is that there be two operands: the first is an input value of type name1 and the output is the name of a variable of the type name2. Name1 is a value on the appropriate stack; it is converted to a value name2, which is placed on the appropriate stack.

As an example,

$$\texttt{int} \rightarrow \texttt{float} \quad 3$$

would take 3 from the argument stack and put 3.0 on the floating point stack.

Files may not be converted to any other type.

# PART I

## MILESTONES

In keeping with the problem-based approach of the text, the text of the milestones is presented first. Chapter 2 presents the specifications of the project. Milestone I asks you to learn a completely new (for most readers) language called Forth. Forth has an interesting history and is currently used to support Adobe's Portable Document Format® (pdf) processing. Forth is the target language for the compiler you will develop. The material in Part II is provided to explain how to develop such a compiler.

Milestones II and III are the first two components for the compiler. The purpose of these two is to read the users' programs, with two results: (1) making a determination as to whether the program is syntactically correct and (2) producing an intermediate, in-memory (deep structure) version of the program.

Milestone IV inspects the deep structure version of a program and enforces the typing rules.

Milestone V produces code for programs that are only constants. While not very interesting in the scheme of things, this is a mid-point check that the compiler is actually capable of producing code.

Milestone VI is perhaps the most difficult of the milestones in terms of implementation. The good news is that once the scope and local variable issue are worked out, the remaining milestones are much easier.

Milestone VII produces user-defined functions. Recursion by itself is not that difficult; the unprepared student, however, often has issues with recursion that must be worked out.

Milestone VIII tackles the issues of nonatomic data definitions. While strings were implemented in Milestone V, the design of strings was under control of the designer. Complex data types take that control from the compiler and place it with the programmer.

Milestone IX is actually an anticlimax: objects are a combination of Milestones VI and VIII. Be sure, however, to read Chapter 3.

# 3

## GENERAL INFORMATION

This chapter provides standing instructions concerning the course. Most importantly, the grading scheme is explained.

### 3.1 GENERAL INFORMATION ON MILESTONES

#### 3.1.1 Contract Grading

Contract grading is a simple concept: the exact criteria for a certain level of grade is spelled out beforehand. This course has milestones, participation, and a final exam. For a one-semester course, the contract matrix is shown in Figure 3.1. This contract scheme is dictated by the highest milestone completed. That is, if a student completes Milestone VI, attends every class, and gets 100 percent on the final, the student still can make no better than a C. Notice that there are no in-term exams; the milestones are those exams.

The milestones must be completed in order. The final grade on the milestones for final grade computation is predicated on the highest milestone completed. The milestone grading scheme is the following:

- The milestone materials will be handed in electronically. The materials must contain a makefile with four targets: `clean`, `compile`, `stutest.out`, `proftest.out`. A skeleton makefile is provided as an appendix. To receive any credit at all for the milestone, the targets `clean`, `compile`, and `stutest.out` must run to completion with no unplanned errors. If these three targets complete, the student receives a provisional score of 50.
- The `stutest.out` target is used to demonstrate the student devised tests. The score is the perceived completeness of coverage. The score on the coverage ranges from 0 to 25 points. The input must be in `stutest.in`. The `proftest.out` target may be used; if the student has modifed the makefile and the `proftest.out` does not work, the student can receive a maximum of 12.5 points.

| | Grading Contract | | |
|---|---|---|---|
| Letter Grade | Milestone[a] | Participation | Final Exam |
| A | Milestone VII | 90% | 90%[b] |
| B | Milestone VI | 80% | 80% |
| C | Milestone V | 70% | 70% |
| D | Milestone IV | 60% | 60% |

[a] Passing score + milestone report.
[b] Exemption possible if done on time.

**Figure 3.1 Grading Example. Passing Grades Are Determined by Milestone**

■ Each milestone is accompanied by a written report based on the `Personal Design and Development Software Process`. Each student must hand in the required PDSP0 forms (plus a `postmortem`). The postmortem report must address two distinct issues: (1) what was learned by the student in designing and implementing the milestone and (2) what was learned by the student about her or his software process. The postmortem is graded on a 25-point scale. *The written report is not handed in electronically, but on paper to facilitate comments on the report.*

## 3.2 MILESTONE REPORT

The milestone project report form is used in the reporting of personal design and software process information outlined in Chapter 11. Each milestone will have slightly different documentation requirements, outlined for each milestone, because we are developing a process. In general, though, the milestone project report will contain three types of information: design documents, project management information, and metacognitive reflective write-up.

### 3.2.1 Design Information

The purpose of the design information is to convince the instructor that the student wasn't just lucky. By this, I mean that the student designed the milestone solution using some rational approach to solving the problem. Because the work breakdown structure (WBS) is an immediate requirement, at a minimum the student's report should include a sequence of WBS sheets. As these sheets are in Microsoft® Excel, there is no particular burden to upgrading them; however, the student needs to show the progression of the design.

### 3.2.2 Software Project Management

The second set of data is that pertaining to the management of the software development process using the spreadsheets provided and described in Chapter 11.

### 3.2.3 Metacognitive Reflection

When we set about assessing the level of thinking that a student is using, we first think about *knowledge*. However, there is more to learning than knowledge—there's general cognitive use. The levels of application are (lowest to highest) comprehension, application, analysis, synthesis, and evaluation. These levels are collectively known as "Bloom's Taxonomy" (Bloom 1956). Higher-order thinking skills refer to the three highest levels; metacognition refers to higher-order thinking that involves active control over the cognitive processes engaged in learning.

Metacognition plays an important role in learning. Activities such as planning how to approach a given task, monitoring comprehension, and evaluating progress toward the completion of a task are metacognitive in nature. Because metacognition plays a critical role in successful learning, learn how to use metacognition in your professional life. While it sounds difficult, metacognition is just "thinking about thinking."

In the milestone report, the student describes the processes that were used in developing the milestone. This includes saying what the student did right and what the student did wrong, or at least could have done better. If the student feels that he or she did something innovative, then the student should analyze that activity in the report. During grading, I look for the student's discussion of "what I learned in this milestone."

## 3.3 GENERAL RESOURCES

- Makefile prototype as in Figure 3.2
- ANS Forth Reference Manual, provided with text
- Gforth Implementation Manual, provided with text
- Information about designing large projects (see Chapter 8 on design)

## 3.4 MAKEFILE PROTOTYPE

The makefile serves two very important purposes. For the student, it guarantees that only the latest versions of the code are included in any test. It also plays an interface role between the student and the grader: the grader does not have to know the inner details of the program. I assume the

```
# Created by D. E. Stevenson, 29 Jan 02
# Modification History
#     Stevenson, 17 Sep 02, added multi-
ple tests to studtest.out
#----------------------
#INSTRUCTIONS
# Quoted strings must to altered by the student
CCC = "compilername"
CCFLAGS = "compiler flags"
OBJECTS = "list of .o files"
SOURCES = "list of source files, including input tests"
RUNFLAGS = "information for runtime flags"
####################################################
#         Warning
# The spaces bfore the commands are not spaces;
# it is a tab. At prompt, type "man make"
# for more information.
####################################################
clean:
 rm -f ".o/class files" core "executables" "outputs"
ls

compiler: clean $(OBJECTS)
$(CCC) $(CCFLAGS) -o compiler $(OBJECTS)

# You may need to specify target/source
information not in the default rules.

stutest.out: compiler
#
# Here is an example of how multifiles can be run.
# Suppose you have several input files, in1, ...
#   and the compiler produces out1, ...
cat in1
-compiler $(RUNFLAGS) in1
cat out1
# the minus sign tells make to ignore the return code.
cat in2
-compiler $(RUNFLAGS) in2
cat out2
# This is the proftest entry. must be there.
proftest.out: compiler
cat$(PROFTEST)
compiler $(PROFTEST)
cat proftest.out
```

**Figure 3.2   Makefile  Prototype**

student has learned how to make a Makefile but I prefer to hand out one prototype.

The makefile has four targets:

**clean:** The clean target removes all work files as well as all previous outputs. The clean ends with listing the directory.

**compile:** This target compiles the milestone code and produces a runnable version of the project to date.

**stutest.out:** The stutest.out target runs the code with the student-prepared input. The student must print out the input test file(s)—there could be several—and the output produced. There should be the same number of output files as input files and each must be clearly labeled.

**proftest.out:** This target allows the grader to test special cases without having to change the student's makefile.

# Milestone I

---

## LEARNING A NEW LANGUAGE, Gforth

---

### Reference Materials

ANSI/IEEE X3.215-199x. Draft Proposed American National Standard for Information — Programming Languages — Forth. X3J14 dpANS-6—June 30, 1993.

### Online Resources

www.gnu.org: GNU Foundation. *Gforth Reference Manual*
www.Forth.org: a wealth of tutorials

---

## MILESTONE REQUIREMENTS: INTRODUCTION TO Gforth

The output of our compiler will be the input to a "machine." In C, the compiler generates an object file, the familiar .o file, that is then linked into the a.out file. To actually run the program you must still cause the operating system to load the a.out file into memory for the machine to execute.

For our project, we will use the program Gforth. Gforth is almost completely the opposite of Lisp. The syntactic format for Gforth is in *Polish Postfix* or, more commonly, *postorder*. Although Lisp has some syntax, Gforth only has spelling rules. As an example, a 1 + 2 expression in C would be entered as [+ 1 2] in SOL and 1 2 + in Gforth.

**SPECIAL INSTRUCTIONS.** Submit your answers in a file named "stutest.in." This makes the filenames consistent across milestones.

Before attempting to do the milestone assignment, read and do the cases contained in the sections entitled "Introduction to Gforth," "Programming in Gforth," and "Introduction to Milestone Cases."

## OBJECTIVES

**Objective 1:** To introduce you to Gforth. Gforth will be the machine that we code to. Typically, students are inexperienced in assembler-level coding. However, to understand compilers and interpreters one must understand how to talk to the underlying machine.

**Objective 2:** To emphasize the crucial role of generalized trees and generalized postorder traversal.

**Objective 3:** To get you to formulate a generalized expression tree data structure.

## PROFESSIONAL METHODS AND VALUES

The professional will learn to use many programming languages and paradigms throughout her or his professional career. The hallmark of the professional in this venue is the ability to quickly master a new programming language or paradigm and to relate the new to the old.

## ASSIGNMENT

The exercises below are simple, "Hello World"-type exercises. These short program segments or small programs are designed to help you to understand how to learn a new language: you just sit down and try some standard exercises that you already know must work.

## PERFORMANCE OBJECTIVES

In this milestone, you have several clear performance objectives.

1.  Either learn to run Gforth on a departmental machine or how to install and use Gforth on your own machine.
2.  Learn the simple programming style of Gforth.
3.  Translate an infix style expression to an expression tree.
4.  Do a postorder traversal of the expression tree to generate the Gforth input. The output of this step is Gforth code.
5.  Produce running Gforth code that evaluates the programs equivalent to exercises. The output here is the running of the Gforth code.

## MILESTONE REPORT

Your milestone report will include versions of items 3 and 4 above. Submit the Gforth input file and makefile electronically. The grader will generate the output file by running Gforth. The standard report format is followed for the paper version of the report.

Your milestone report must include a data structure for an n-ary tree and a pseudo-code recursive algorithm to translate an arbitrary instance of these trees into postorder.

## FORTH EXERCISES

The required material for this milestone consists of two sections. One section is an exercise to translate C code snippets into Gforth code. The second section is an exercise in Gforth to implement string support needed in Milestone V; experience shows that this string exercise is a valuable learning tool.

1.  Pseudo-C exercise. The following are statements in "pseudo-C." Write Gforth programs that will compute the same thing the C program would using the statements in Figure I.1
2.  Gforth strings. Gforth's strings are even more primitive than those of C because the fundamental convention of the *delimited* string is absent. Recall that in C, a "string" is a one-dimensional char array that is terminated by the null value. The delimited approach is only one such convention: we could include the length of the string at the beginning for what is termed a *counted* string.

## CASE 1. STRING REPRESENTATION

Develop *your* representation for strings. The choice between delimited and counted should be made based on the trade-off between representation size and ease of operation. This decision is primarily based on *concatenation*. For this milestone, implement concatenation in your choice of representation.

## INTRODUCTION TO Gforth

Charles Moore created Forth in the 1960s and 1970s to give computers real-time control over astronomical equipment (see Moore 1980). Functions in Forth are called "words." The programmer uses Forth's built-in words to create new ones and store them in Forth's "dictionary." In a Forth program, words pass information to one another by placing data onto (and removing data from) a "stack." Using a stack in this way (Forth's unique contribution to the world of programming languages) enables Forth applications to run quickly and efficiently. The use of stacks in this way is familiar to users of scientific hand calculators such as the Texas Instruments TI-86.

| Problem | Statement |
|---------|-----------|
| 1 | `printf("Hello World\n");` |
| 2 | `10 + 7 - 3*5/12` |
| 3 | `10.0 + 7.0 - 3.0*5.0/12.0` |
| 4 | `10.0e0 + 7.0e0 - 3.0e0*5.0e0/12.0e0` |
| 5 | `10 + 7.0e0 - 3.0e0*5/12` |
| 6 | `y = 10;` |
| | `x = 7.0e0;` |
| | `y + x - 3.0e0*5/12` |
| 7 | `if 5 < 3 then 7 else 2` |
| 8 | `if 5 > 3 then 7 else 2` |
| 9 | `for ( i = 0; i <= 5; i++ )` |
| | `printf( "%d ",i);` |
| 10 | `double convertint(int x){ return ((double)x);}` |
| | `convertint(7)` |
| 11 | `int fibonacci(int i) {` |
| | `if (i < 0 ) abort();` |
| | `else if (i == 0 ) return 1;` |
| | `else if (i == 1 ) return 1;` |
| | `else return fibonacci(i-1) + fibonacci(i-2);` |
| | `}` |

**Figure I.1    Milestone I Elements**

The Forth Interest Group* and the Institute for Applied Forth Research[†] help promote the language. Two books by Brodie (1981, 1984) are perhaps the best-known introductions to Forth. Perhaps the most famous application is as Adobe's PDF® (portable document format).

Forth went through several standarization cycles. The final cycle ended in 1994. The Proposed Standard is available online from several vendors.* There are many implementations of Forth available; the Forth Interest Group lists 23 commercial vendors as of April 2005. Several implementations are open source. The implementation discussed here is available from the GNU Project server at www.gnu.org.

A Google search of the Internet returns many tutorials on Forth. Many of these tutorials are for a specific implementation although almost any such tutorial presents a general introduction. Any Forth tutorial after the 1994 Standard should contain information consistent across the spectrum. The Gforth manual accompanying the implementation contains an excellent tutorial.

---

* http://www.forth.org
[†] dec.bournemouth.ac.uk/forth/rfc/inst.html
* http://www.taygeta.com/forth/dpans.htm being just one.

## PROGRAMMING IN Gforth

The reader is certainly familiar with expressing arithmetic calculations in a language like C or Java. For example, consider the following:

$$1 + 2 * 3$$

which is evaluated to 7. Such a presentation is called *infix notation*. But this is not how such a calculation is expressed to the machine.

True assembler languages reflect the operational nature of the underlying machine's machine language. The distinction between *machine* language and *assembler* language is primarily the medium of presentation: machine language is electronic and assembler is character. Assemblers were developed because it was too difficult and error-prone to use machine language for large problems. Assemblers generally include features that make programming easier for the human but which must be translated (usually in a transparent manner) to machine information.

A generic picture of an assembler statement is the following:

$$\text{operation operand}_1, \ \ldots, \ \text{operand}_n$$

The operation could be a primitive data operation, a primitive control operation, or the name of a programmer-defined function. The operands could be primitive data constants or expressions relating to variables or addresses. Presented this way, we call such a notation *prefix notation*. Operations can be categorized by how many operands can be mentioned in a statement; that is, how large $n$ is in the schema above. The above $1 + 2 * 3$ calculation might be denoted in assembler as below:

```
load  r1, 2
mult  r1, 3
add   r1, 1
```

which computes 7, leaving the result in a place called r1.

Forth uses no addresses directly but relies on a stack to hold all values. Such a system results from a *postfix notation* often used in scientific handheld calculators. The postfix notation for our calculation might be

$$1 \ 2 \ 3 \ \textit{mult plus}$$

leaving the 7 on the top of the stack. Postfix-based processors are very fast as they bypass much of the fetch-execute cycle. Notice that no intermediate assignments are required to store temporary values in this postfix execution style.

## INTRODUCTION TO MILESTONE CASES

The goal of this milestone is to master some of the fundamentals of the Gforth. The fundamental processes to master in this milestone are

1. Take representative programming expressions in the C family and put them into a tree.
2. Process the tree in a postfix fashion to design the Gforth program.
3. Translate the design into Gforth.
4. Write the Gforth program.
5. Test the Gforth program.

## CASE 2. ELEMENTARY ARITHMETIC STATEMENTS

The first case study touches on some issues that you are already familiar with from your data structures class. Compiling is based on three basic ideas that you learned in algebra, although you were not taught with these words:

1. Every arithmetic expression can be put into a tree.
2. Every tree can be traversed in postorder.
3. The list generated by the postorder traversal can be evaluated to produce a value without intermediate storage.

For each of the expressions in Figure I.1 below,

1. Draw the tree that represents the correct precedence.
2. From the tree, generate a list by a postorder traversal.
3. From the postorder traversal list, generate the value.
4. From the postorder traversal list, write the Forth program, run it, and compare the value computed by a hand calculator.

### Assignment

Part 1. Evaluate the arithmetic statements in Figure I.2. **Caution:** these statements are presented as someone might write them in a mathematics class and not in a computer science class. This introduces the concept of *context* or *frame of reference*. In particular, carefully consider the values of the last two expressions before you compute.

Part 2. Run the resulting code in Gforth.

Part 3. On one page only, write the algorithm that takes any *infix* statement into a *postfix* statement.

$$(1) \quad 1$$
$$(2) \quad 1+2$$
$$(3) \quad 1+2 \times 3$$
$$(4) \quad -1+2 \times 3$$
$$(5) \quad -1+2 \times 3^2$$
$$(6) \quad 1+2 \times (-3)^3$$
$$(7) \quad -1+2 \times 3.0^{4.7}/6$$
$$(8) \quad 2^{-3^3}$$
$$(9) \quad 2^{-5}$$
$$(10) \quad \max\{1, 2, 3, 4, 5\}$$

**Figure I.2  Case 1 Problems**

## CASE 3. ROUND 2: MORE ARITHMETIC

The second case study touches on more complex issues in evaluation, but still working with just arithmetic expressions. We want to add variables. We still want

1. Every expression to be put into a tree.
2. Every tree traversed in postorder.
3. The list generated by the postorder traversal to be evaluated to produce a value without intermediate storage.
4. From the postorder traversal list, write the Forth program, run it, and compare the value computed above.

As with Case 1, the rules are

1. To take the mathematical expressions below and draw the tree that represents the correct precedence.
2. From the tree, generate a list of the leaves by a postorder traversal.
3. From the postorder traversal list, generate the Gforth input.
4. Test the resulting Gforth expression.

After you have done the exercises, update the algorithm you wrote for Case 1 to reflect what you have learned.

In Figure I.3 we want to consider sets of expressions, because to have variables, we need to have definitional expressions as well as evaluative expressions.

This is a significantly harder exercise than Case 1. Some of the issues are

1. How are you going to handle the semicolon?
2. How do you declare a variable in Gforth. The declaration sequence does not obey the strict postfix rule. Why?

$$\{x = 1; x\}$$
$$\{z = 3; 1 + z\}$$
$$\{y = 10; -1 + y \times 10\}$$
$$\{y = 3.0; 1 + 2 \times y\}$$
$$\{x = 3; y = 7; -1 + x \times 3^y\}$$
$$\{q = 3; 1 + 2 \times (-z)^3\} \quad \text{This is correctly written}$$
$$\{e = 4.7; -1 + 2 \times 3.0^e/6\}$$
$$\{a = 3; b = -3^a; 2^b\}$$

**Figure I.3    Sets of Expressions**

3. How do you assign a value to such a variable?
4. How do you retrieve the assigned value?
5. What happens if an undeclared variable is evaluated?

Run the resultant codes in Gforth.
Revise your algorithm produced in Case 1.

## CASE 4. IF AND WHILE STATEMENTS AND USER-DEFINED FUNCTIONS

Forth was designed as an interactive language and is probably quite different than what you're used to. Most importantly, Forth has two different input modes: interactive and compile. Words that are evaluated generally cannot be used in the compile setting and vice versa. This leads to the rather strange situation of having two if statements: one in the interactive mode and one in the compile mode. Each has its place, but we will concentrate on the compiled if.

Make up test cases to demonstrate the use of if and while statements. Be sure you understand the differences between the two modes. Run the test cases to demonstrate your understanding.

### Defining User Words

In Chapter 6 we introduce the concept of the metalanguage that enables us to talk *about* a language. We make use of that concept here. In Forth, words are defined in terms of a sequence of words and constants.

In order to discuss a language, for example Gforth, using another language, for example English, we must use the idea of metalanguage and object language. In our case, Gforth is the object language and English is the metalanguage. This gives rise to a convention concerning meta notation.

We will use the angle brackets around a symbol to indicate meta-names. In this case, ⟨*Statements*⟩ means "zero or more statements."

We use the term ⟨*ForthCompilableExpresssion*⟩ to mean any legitimate sequence of Forth words except a colon and a semicolon and ⟨*New Word*⟩ to indicate the word being defined. The syntax for a user-defined word in Forth is

: ⟨*NewWord*⟩ ⟨*ForthCompilableExpression*⟩ ;

Remember that you need blanks surrounding the colon (:) and the semicolon (;).

### *if* and *while*

Now we're able to use `if` and `while` statements as they normally appear in the C family. Using metavariables in the obvious way, we can define `if` and `while` as follows:

⟨*Condition*⟩ if ⟨*TrueComputation*⟩ else ⟨*FalseComputation*⟩ endif
⟨*Condition*⟩ while ⟨*Computation*⟩ endwhile

This case explores the two control structures, `if` and `while`. Here the postfix regime breaks down even further. We still can put the expressions into a tree, but we cannot evaluate everything: we can only evaluate the proper statements.

## CASE 5. COMPLEX DATA TYPES

Complex data types such as strings and arrays, C `structs`, and Java objects must be developed if the project is to be able to emulate the full range of data types that can be defined using the type systems of modern languages. In a one-semester course at the introductory level, it is not possible to develop a compiler that can reliably compile structures, objects, and the referencing environment.

However, it is possible to develop support for character strings. For many, this will be quite a challenge because it requires implementing several complex functions. One design approach is to use the CRUD (create, retrieve, update, delete) approach. This is a commonly used approach to database design. We return to this issue in Milestone V.

## CASE 6. FUNCTIONS

The third case study touches on the programming (rather than the mathematical) context issues in evaluation, but still working with just numbers. We need to add local variables and definitions. The goal remains the same: investigate the algorithmic issues in evaluation.

Below, we want to consider sets of expressions, because to have variables, we need to have definitional expressions as well as evaluative expressions.

```
define n! = if n <= 0 then 1 else n*(n-1)!
```

Here, the issue is much different than the straight arithmetic problems in Case 1 and Case 2. Now we have to make multiple copies and evaluate those copies. This process is called *unfolding*.

We have now seen an example of all the straight computational and evaluation issues.

Run your encoding for 5!.

# Milestone II

---

# LEXICAL STRUCTURE
# AND SCANNER

---

**Reference Materials**

Terrence W. Pratt and Marvin V. Zelkowitz. *Programming Languages: Design and Implementation*. 4th Ed. Upper Saddle River, NJ: Prentice Hall. 1995. Section 3.3.

Kenneth C. Louden. *Programming Languages: Principles and Practice*. 2d Ed. Thompson, Brooks/Cole. 2003. Section 4.1

---

This milestone begins a sequence of four milestones (Milestone II–Milestone V) that use the same basic approach to design and implementation. In each of these milestones, we begin by using a formal language. The formal language is used to capture the solution of the milestone problem. Each formal language solution specifies an abstract program that can then be encoded into the implementation language.

The goal of this milestone is to design a *scanner*, the enforcer of the lexical definition of SOL. The purpose of a scanner is to read the individual characters from the input file and convert these characters in a token. *Tokens* are data structures that carry (1) information read by the scanner and (2) the terminal classification according to the grammar rules. The specification for the scanner comes from the *lexicon* as described by the system specification (Chapter 2).

This milestone also begins development of the *symbol table*. The symbol table is developed throughout the project.

## CASE 7. PROJECT MANAGEMENT

Before beginning this milestone, read Chapter 11. This chapter describes elementary project management processes for this project. The central focus should be the development of the WBS spreadsheet. Before attempting any milestone, the WBS relevant to that milestone should be included in the project management portfolio.

In order to start at a common place, Figure II.1 contains a skeleton WBS. Using this WBS, develop your own based on the cases in each milestone.

## OBJECTIVES

**Objective 1:** To develop a formal definition of the scanner algorithm from the lexical structure for SOL. You will code the scanner from this formal definition.

**Objective 2:** To formalize the data structure used for the tokens.

**Objective 3:** To give you experience using a *state-transition diagram* as a design tool.

**Objective 4:** To give you experience in designing a program using a categorical style design.

**Objective 5:** To begin work on the symbol table for the project.

**Objective 6:** To give you practice in designing test files based on the formal definition and develop a test driver for the combined scanner-parser program.

**Objective 7:** To test the resulting program for correctness based on the formal definition.

**Objective 8:** To develop personal project management skills.

| Task Number | Description | Inputs From | Outputs To | Planned Value | |
|---|---|---|---|---|---|
| | | | | Hours | Cum. Hours |
| 1 | MAIN Function | | | | |
| 1.1 | Initialize | | | | |
| 1.2 | Open files | | | | |
| 1.3 | Call Parser | | | | |
| 1.4 | Call Type Checker (stub) | | | | |
| 1.5 | Call Semantics routines (stub) | | | | |
| 1.6 | Call Code Generator (stub) | | | | |
| 1.7 | Close files | | | | |
| 2 | Tree Data Structure | | | | |
| 3 | Token Data Structure | | | | |

**Figure II.1    Beginning Work Breakdown Structure (WBS) Based on Requirements**

## PROFESSIONAL METHODS AND VALUES

This milestone is the first of two that use formal language theory to specify the program to be written. In particular, scanners rely heavily on the *finite state machine* (FSM) formalism.

Because the formalism allows for complete specification of the lexicon, test case development is relatively easy. The grading points for this milestone are based on the fact that you should be able to test all cases.

## ASSIGNMENT

The purpose of this milestone is to develop a working scanner from the lexical definitions. Scanners are supported by over 50 years of development; these techniques were used by engineers in the 1960s to design circuits. Sections entitled "Scanner Design" and "Scanner Design Extended" provide details.

This is the proper time to start the symbol table development. The only requirement is that the lookup routine be an open list hash table: the lookup must work in $O(1)$. See the section entitled "Symbol Tables" for details.

## SPECIAL REPORTING REQUIREMENTS

Starting with this milestone, each milestone report will be accompanied by all design documentation along with your PDSP0, WBS, and time recording template (TRT) forms. The milestone report must include your FSM design, as well as your metacognitive reflection on what you did in the milestone and what you learned.

## ADDITIONAL INFORMATION

The lexical scanner rules are the same as C whenever a specific rule is not stated. For example:

1. Specialized words in SOL that are the same as C take on their specialized meaning. For example, the `if` and `while` have approximately the same semantics in SOL and in C.
2. Direct information on the format of integers has not been given, so C rules are used; the same is true for floating point and strings. **Caution:** Floating point numbers are somewhat difficult to deal with terminologically. SOL and C have the same view of floating point values; Gforth does not share that definition. In Gforth, 3.2 is treated as a `double` integer when 3.2e is treated as `floating`

point; SOL has no double but C's double is equivalent to Gforth's floating point. Confusing? No, the context keeps it straight.

# SCANNER DESIGN

The scanner reads in the program text and accumulates characters into words that are then converted into tokens. The scanner operates as a sub-routine to the parser. The communication between the two is in terms of instances of tokens. The purpose of the token structure is to inform the parser of the terminal type of the token and the value associated with the token.

Conventionally, the scanner is designed using FSM concepts. The work of the cases is to design that machine. The possible tokens are defined by the lexical rules of the language defined in Chapter 5.

# CASE 8. DESIGNING THE TOKENS

Read the specification documents and the parser and determine the various token classes that are required. Now develop a data structure for these tokens. Tokens are a data structure that contain both the "word" read from the input and an encoding of the terminal type. The word must be stored in its proper machine representation; that is, integers are not stored as strings of characters but in their 32-bit representation. **Requirement:** If you are using C, then you must define a *union;* in object-oriented systems you will need an *inheritance hierarchy.* You will develop some of the information for this case in Case 47.

Realize that the scanner is required to check constants for validity. As an example, a sequence of digits 20 digits long cannot be converted to internal representation in 32 bits. Therefore, this 20 digit number is illegal. Hence, the words are converted to their proper memory format for integers and floating point numbers.

## Scanner Design

### Finite State Automata

The finite state automata (FSA) (sing. automaton) paradigm is the basis for the design of the scanner. The FSA paradigm as described in theory texts has two *registers* (or memory locations) that hold the *current character* and *current state.* The program for the machine can be displayed in several standard ways. The graphical representation is easier to modify during design. The implementation program that is the scanner is written using the graph as a guide.

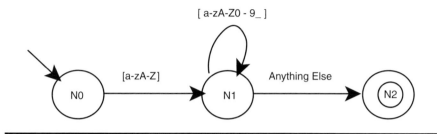

**Figure II.2    Finite State Machine, Graphic Presentation**

The graph in Figure II.2 represents a program that recognizes a subset of C variables. State $N0$ is the start state. If the machine is in the start state and it reads any upper- or lower-case alphabetic chararacter, the machine moves to state $N1$; the notation used in the diagram is consistent with Unix and editor notations used for *regular expressions*. As long as the machine is in state $N1$ and the current input is any alphabetic letter, a digit, or an underscore, the machine remains in state $N1$. The second arrow out of state $N1$ is the terminating condition. It should be clear with a moment's reflection that the self-loop on $N1$ is characteristic of a while statement. With that observation in mind, the C code in Figure II.3 can be used to recognize variable names. By convention, the empty string symbol $\epsilon$ means that no character is needed for this transition.

To understand the programs, we must first understand the machine execution cycle. This is actually not much different from the cycle used in a standard computer.

> **MC1:** The machine is in a state as indicated by the current state register. If that state is a valid state, then continue; otherwise go to step 5.
>
> **MC2:** Read the next character from the input. If there is no next character, halt and go to MC5.
>
> **MC3:** The program of the machine is a set of pairs *(state, character)*. Search the list of valid pairs for a match with the contents of the current state, current input registers.
>
> **MC4:** If there is a match, set the current state to the value associated with (current state, current input) and go to MC1. If there is no match, set the current state to an invalid value and go to MC5.
>
> **MC5:** The machine has a list of special state names called final states. If, upon halting, the current state value is in the final state list, then the input is accepted; otherwise, it is rejected.

The theory of FSA describes what inputs are acceptable to finite state automata. To develop the theory, consider a formalization by asking what

```
#include <ctype.h>
#include <stdio.h>

void recognize_name() {
    int current_input;
    enum state { N0, N1, N2 };
    enum state current_state = N0;
    current_input = getchar();
    if(current_state == N0 && isalpha(current_input))
      current_state = N1;
    current_input = getchar();
    while( current_state == N1 &&
            (isalpha(current_input) ||
             isdigit(current_input) ||
            current_input == '_')) {
      current_state = N1;
      current_input = getchar();
    };
    current_state = N2;
}
```

**Figure II.3    C Program for FSA Graph**

the various parts are. There are five:

1. The set of possible characters in the input (Programming-wise, this is the type of input and of the current input register.)
2. The set of possible state values in the current state register
3. The full set of pairings of *(state, character)*
4. The list of final states
5. The start state

Part of the difficulty of studying such abstractions is that the formulations can be quite different and yet the actual outcomes are the same. For example, this same set of concepts can be captured by *regular expressions, regular production systems,* and *state transition graphs.* The regular expressions used in editors like *vi* and pattern matching programs like *grep* are implemented by an automaton based on the above definition.

For design purposes, the preferred graphical representation is that of a *state-transition graph (ST graph)*. A state-transition graph is a representation of a function. In its simplest incarnation, an ST graph has circles representing states and arrows indicating transitions on inputs. Each of the circles contains a state name that is used in the current state. The tail of the arrow represents a current state, the character represents current input, and the head of the arrow names the new state.

ST graphs are representations of functions. Recall that a function has a domain and a range. In the ST graph, the domain and range are circles and we draw an arrow from the domain to the range. The arrow is weighted by information relating to the function. The ST graph is directed from the domain to the range. It is customary in finite state automata work to have an arrow with no domain as the start state and double circles as final states. If every arrow emitting from the domain circle is unique, then the transition is *deterministic*; on the other hand, if two arrows weighted the same go to different ranges, then the transition is *non-deterministic*. It is possible to write a program that emulates either a deterministic or non-deterministic automaton; for this milestone, let's think deterministically.

The weight of the arc captures the functioning of the transition. Since FSA only need to indicate which character is involved, the graphs in texts are very simple.

## CASE 9. DEVELOPMENT OF FINITE STATE AUTOMATA DEFINITIONS

Develop a finite state automaton for each class of lexical input specifications.

### Finite State Machines

The technical difference between an *automaton* and a *machine* is that an automaton only gives an accept-reject response, whereas a machine does transformations.

Case 9 only develops the accept-reject portion of the scanner. The scanner must produce a token as output and, therefore, we need to think about the saving of characters, possibly converting some (like integers or special characters in strings) etc. We denote this by the weight. Figure II.4 illustrates the concept. That figure is just Figure II.2 with the function to be performed *after* the pattern is matched. Finite state machine diagrams are great design tools.

## CASE 10. FINITE STATE MACHINE DESIGN

Modify your output of Case 9 to reflect the processing to save and convert characters to form tokens.

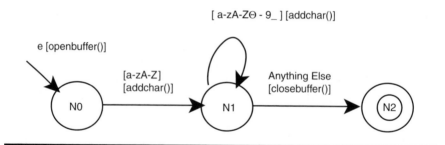

**Figure II.4   Finite State Machine, Graphic Presentation**

## Guidance

The purpose of all these exercises is to show you a professional method often used in hardware, protocol, and input design.

## SCANNER DESIGN EXTENDED

We now have a first design of our scanner. We need to now consider whether or not the design is worth anything. We will do this with two exercises.

### The State Exercise

Unfortunately for those who need to use the FSA and FSM ST graphs to design things, the graphs as the theoreticians have them are not completely useful. For any graph of any complexity, we need to ask, "What is the semantics of this graph?" That is, what does this graph actually do?

In theoretical studies, the states are treated as a more or less meaning-less alphabet: a set of letters, say. That's actually not very interesting. What is interesting is that the states are predicates. *Predicates* are statements in which something (the *predicate*) is affirmed or denied of something (the *subject*), the relation between them being expressed by the *copula*. The ex-tended definition means the form of thought or mental process expressed by this, more strictly called a *judgment* (*Oxford English Dictionary*). Inter-estingly enough, judgments can always be displayed as a directed acyclic graph (DAG).

## CASE 11. REASONING ABOUT DESIGN

For each of your ST graphs, annotate the nodes with predicates concerning the computation. Write down a list of test cases that can be used to show that your ST graph is correct. Using the predicates, choose representative test cases and interpret the ST graph and show that this test case is correct. Keep it Simple!

## WORK BREAKDOWN STRUCTURE

We have designed the scanner and we have seen that it can be used to design tests for the complete scanner. But this is only part of the milestone because we must translate the design into a program in your favorite programming language. To this point, we have been focusing on logical sequence of flow; but there are other issues, such as

1. Opening and closing input files
2. Using error protocols
3. Saving the input string for the token
4. Creating new instances of a token

This is where the WBS exercise is crucial.

In the WBS form (see Chapter 11), the critical exercise is to name every possible task that must be accomplished for the scanner to function. The more complete the decomposition (breakdown), the better. Ultimately, we must be able to estimate the number of lines of code that we will have to write. The hours needed to code a function is the number of lines of code divided by 2.5.

## CASE 12. SCANNER WORK BREAKDOWN STRUCTURE

Produce a work breakdown structure for the scanner. Keep a record of how much time (earned value) you actually spend. Turn the work breakdown structure in with the milestone report.

## SYMBOL TABLES

Symbol tables are the heart of the information management functions in a compiler. At their core, symbol tables are database management systems. We will address the design of symbol tables in this manner.

### Database Management 101

A database is a collection of data facts stored in a computer in a systematic way so that a program can consult the database to answer questions, called queries. Therefore, there are three fundamental issues in the design:

1. What are the queries that the database is meant to provide information about?
2. How is the data retrieved by a query?
3. What is the actual data in the database and how is it turned into information?

A typical database has a basic layout as shown in Figure II.5.

Many different design disciplines have been proposed over the years based on underlying data organization (trees or directed graphs) or query formation, such as the relational approach. These disciplines are often suggested by the principal uses of the database system, the so-called "enterprise" model that must satisfy a divergent user population. We don't have that problem: the compiler is the sole user, so we can design the system to be responsive to a limited set of data and queries.

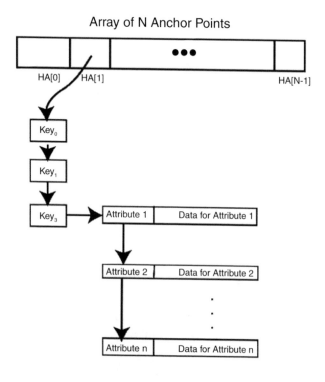

Key - Attribute - Data Paradigm

**Figure II.5   Simple Database Configuration**

## Initial Analysis

One approach to designing a symbol table is to simply ask, "What are the entities that should have data in the symbol table?" An approach to that is to look at the classes of symbols in the lexicon. We basically have symbols that are (1) constants (predefined with a fixed value), (2) tokens that have no semantic import, like '[', and (3) names. Since (2) represents information strictly between the scanner and parser, let's ignore them for the time being.

Constants and names, then, are problematic. What is even more problematic is the fact that until we have specific queries we don't know what data we need. This argues strongly for a very flexible approach.

On the other hand, the basic look-up requirement is clear: we need to be able to look up the entry associated with a "name" (string of characters). This is a data structures issue. The requirement here is for very fast look-up.

## Second-Order Analysis of Symbols

This is a good point at which to review the organization of a natural language dictionary, such as the *Merriam-Webster Unabridged* or the *Oxford English* dictionaries. For example, look up "lexicon" in the *Oxford English Dictionary*, second edition online. If you do, you will find that there are three major definitions, some accompanied by minor definitions. What we learn from this is that the dictionary (symbol table) must be aware of the *context* in which the query is being asked.

As a second experiment, look up the word "party." We find that there are four major definitions, one for different *parts of speech*. What is the equivalent concept in programming languages to "part of speech?" Parts of speech tell us the *use* of a word. Now we have something to go on: the basic organization in the symbol table is that there is one class (structure) of data for every possible use a symbol can have in a program.

From these two examples, we can conclude that we should take an *entity-property-value* approach to the design of the symbol tables. The individual entities are either constants or names. For every possible use that an entity can have, there should be a data structure that holds the unique values associated with the entity for this property.

The symbol table will be developed throughout the project.

## CASE 13. SYMBOL TABLE DESIGN

The generic symbol table has three basic considerations: (1) a hash function to convert *keys* to *integers*; (2) a collision maintenance system; and (3) a variety of data structures, one for each use. A given key may have more than one data structure associated with it.

Design and implement a symbol table system for this milestone. The symbol table for this milestone must identify *keywords* such as let as keywords.

# Milestone III

---

# PARSING

**Reference Material**

Terrence W. Pratt and Marvin V. Zelkowitz. *Programming Languages: Design and Implementation.* 4th Ed. Upper Saddle River, NJ: Prentice Hall. 1995. Section 3.4.

Kenneth C. Louden. *Programming Languages: Principles and Practice.* 2d Ed. Belmont, CA: Thompson, Brooks/Cole. 2003. Chapter 4.

This milestone completes the syntactic phase of the project. The parser directs the syntactic processing of the input and builds a version of the program in memory. The whole syntactic process fails if there is a grammar or lexical error.

In Milestone II we developed the program by using graphic definitions of finite state machines. The graphic representation of FSMs is not powerful enough to describe the processing of context-free languages. Instead, we will generalize another concept, that of *production systems*.

## OBJECTIVES

**Objective 1:** To produce the parser and continue the symbol table for the project

**Objective 2:** To use the data structure developed for the tokens

**Objective 3:** To give you experience using the production system concept as a design tool

**Objective 4:** To give you experience in designing a program using a categorical style design

**Objective 5:** To give you practice in designing test files based on the formal definitions

**Objective 6:** To develop personal project management skills

# PROFESSIONAL METHODS AND VALUES

This milestone requires careful planning in both the design and testing phases. We will use a formal method. Parsing has perhaps the strongest theoretical grounding of any subject in computer science. We review this material in the sections entitled "Grammars and Things Grammatical," and "Parser Design II," on the Chomsky Hierarchy.

# ASSIGNMENT

The purpose of this milestone is to develop a working parser utilizing the scanner and symbol table. Performance objectives are to

1.  Develop a formal definition of the recursive descent parsing algorithm from the grammatical structure for SOL
2.  Develop a parser program from a formal definition
3.  Develop a data structure for the abstract syntax graphs, also known as abstract syntax trees
4.  Develop a test driver for the developed program
5.  Test the resulting program for correctness based on the formal definition

The milestone report should concentrate on the development of the program and what you learned about formal production system design concepts. It must include a copy of your formal design documents.

## Detailed Information

You are to develop the following:

■ A scanner that reads characters from a file or standard input. The file names are taken from the command line. The task of the scanner is to classify the terminal symbols as specified by the parser's grammar.
■ A parser is a program that gets its input from the scanner. The parser has two tasks:
1.  The parser verifies that the input meets the rules of the grammar.
2.  The parser produces a tree that encodes the input.
■ The main program ("driver") sets up the input files by reading the command line flags. You should consider using flags to communicate debugging options. Any flag that does not start with the customary hyphen ('−') or a plus ('+') is taken as a file to be processed. This subsystem is the input user interface and could have multiple

input files. **Note:** A file can have multiple expressions. The loop outlined above must be continued until all files are processed. The process loop is as follows:

1. Set up the file to be processed so the scanner can read it.
2. Call the parser to parse the input and to receive the tree produced by the parser.
3. Echo print the tree to verify the input is correct. The printer program should be a separate, stand-alone function that can be used throughout the compiler's development.

## Testing Requirements

The trees will be used eventually to fulfill future milestones. For this project, you are to demonstrate that your parser can (1) properly process correct inputs and (2) reject incorrect inputs. You want to keep the print subroutine as a debugging aid for later in the project.

## The Project Grammar

The project grammar is given in the specifications.

# CASE 14. DO WE HAVE TO WORK HARD?

One possible approach to the parser is to implement—as directed below—the grammar in the specifications. However, careful thought (and maybe a little research) would indicate that there is a very simple underlying grammar that can be developed. This simpler grammar is a trade-off between what is *syntactic* and what is *semantic.*

If you take this much simpler route, then you must justify this in the milestone report. In particular, you must develop an argument as to how and why you will guarantee that a program that is not correct by the specification grammar will be stopped. **Hint:** Move some syntactic considerations to semantic considerations.

## Specific Grading Points

- ■ Constraints
  - – You are to implement the parser as a recursive descent parser.
  - – The scanner must be a subroutine called by the parser.
  - – You must not store the parentheses in the tree.
  - – You may not cast to Object. Any variant record implementation must either be by inheritance for Java and C++ or unions in C.

- You must be able to read any number of input expressions—properly bracketed lists—made up of an arbitrary number of subexpressions and properly constructed lexical elements.
- You may not use counting of parentheses as the parser algorithm.

■ Conditions. Recovery from input syntax errors is almost impossible. Throw an error and exit.

## TESTING REQUIREMENT

The program will be accompanied by test files. These test files are to be developed by you and should clearly demonstrate that the scanner and parser implement the complete specification. They should clearly show both correct processing of correct inputs and rejection of incorrect inputs.

## GRAMMARS AND THINGS GRAMMATICAL

We will use the vocabulary of *formal languages* to describe formal programming languages and their parsers.

For technical reasons, FSMs such as we used to design the scanner are not powerful enough to deal with context-free programming languages; this is because it is impossible to set a viable limit on the number of "parentheses" in advance. It is necessary to move to a more powerful model, *context-free grammars*. Specification of context-free grammars is more complicated than for FSMs, but the concepts are similar. We are unable to use a simple ST graph for context-free languages; this is due in part to the need to specify recursion.

A grammar $G$ is a four-tuple ($V$, $T$, $P$, $S$), where $V$ is a set of *categorical symbols; T* is a set of *terminal symbols; P* is a set of *grammar rules;* and $S$ is a member of $V$ known as the *sentential symbol.*

In order to have a definite example and to maintain continuity with the case studies, we consider the grammar for arithmetic expressions, known in the "compiler writing trade" as *Old Faithful*, because it is an easy, complete example of the issues.

### Basic Terminology

The motivation for formal languages is natural language, so the initial concepts are all borrowed from natural language. Grammars describe *sentences* made out of *words* or *symbols*. Both $V$ and $T$ are sets of such words or symbols.

A language would be all the sentences that are judged correct by the grammar: this is called *parsing*. Technically, we define the *Kleene closure*

$$
\begin{array}{ccc}
N0 & \rightarrow & a\ N1 \\
N0 & \rightarrow & b\ N1 \\
. & . & . \\
. & . & . \\
. & . & . \\
N0 & \rightarrow & Z\ N1 \\
\end{array}
$$

**Figure III.1  Regular Productions for Recognizing Names**

*operator* $A^*$ to mean zero or more occurrences of a string $A$. This notation is probably familiar to you from the use of regular expressions and as the "wildcard" operator in denoting file names. If $A$ is a set, then $A^*$ is every possible combination of any length. If $T$ is the set of all possible terminal symbols, then $T^*$ is the set of all possible strings that could appear to a language, but not all may be grammatically correct.

Which strings of $T^*$ are the "legitimate" ones? Only those that are allowed by the rules of $P$. So how does this particular description work?

## Productions

### Regular Grammars First

Productions are another way to describe a pattern. The productions are two words, for example, $Y$ and $Z$. Conventionally, a production in $P$ is denoted $Y \rightarrow Z$. Having already learned how to describe patterns with graphs (Milestone II) we can move from graphs to productions. In the graphs, we have nodes and arrows. Each node has a label and each arrow has a weight. We can collect all the node labels into a set; call this set the *nonterminals*. Likewise, we can collect the set of individual weights in the *terminal* set. A production is essentially a rule taking one nonterminal to another while recognizing input terminals.

As an example, consider the FSA developed to recognize names (Figure III.1). The start node is $N0$. We can traverse to $N1$ if the regular expression $[a - zA - z]$ is satisfied. In words, "Start at $N0$. If any alphabetic character is on the input, the next state is $N1$." The same concept is captured by the productions in Figure III.2.

## CASE 15. CONVERSION OF FINITE STATE AUTOMATON GRAPHS TO GRAMMARS

Complete the conversion of the "Name" graph from its graphic form to production rules.

$$E \rightarrow T + E$$

$$E \rightarrow T - E$$

$$E \rightarrow T$$

$$T \rightarrow F * T$$

$$T \rightarrow F / T$$

$$T \rightarrow F$$

$$F \rightarrow \text{any integer}$$

$$F \rightarrow (E)$$

**Figure III.2    Productions for Arithmetic Expressions without Variables**

Convert the floating point number definition from its FSM (not FSA) graphic form to a production form. **Hint:** Think of the FSA graph as an FSM graph that has no actions.

### Context-Free Productions

The regular language basis of the scanner is deceptive. Despite its apparent power, it has a flaw when it comes to describing languages such as SOL. Regular languages cannot balance parentheses because they have no memory. The strict form of the productions $N \rightarrow tN'$ with $N$ and $N'$ being nonterminals and $t$ being a terminal matches the fact that the FSA have only current input and current state memory locations. In order to recognize balanced parentheses, we need more memory; in fact, a simple stack will do. This is the model of *context-free languages*.

In order to deal with the parentheses issue, we need to loosen the rules on what constitutes a legitimate production. In order to have a definite example, we consider the definition of Old Faithful (Figure III.2), so named because it appears in most language texts; it captures the essence of arithmetic expressions.

To begin, we have two sets of symbols (just as in the above regular example): $V$ is the set of nonterminals and $T$ is the set of terminal symbols.

$$V = \{E, T, F\}$$

$$T = \{+, -, *, /, (, ), \text{any integer}\}$$

The members of a new set, call it $P$, are the productions. These rules describe an algorithm by which we can decide if a given sequence of symbols is grammatically correct.

**Note:** There is a contextual use of the symbol $T$ in two different ways: $T$ in the definition of grammars as the terminal set and $T$ in our example grammar (short for *term*). Such overloading of symbols is common; there are only so many letters. Therefore, you must learn to be cognizant of the context of symbols as well as their definition in context. Ambiguity should arise only when two contexts overlap.

## Derivations

A sentence is structured correctly if there is a *derivation* from the starting symbol to the sentence in question. Derivations work as shown in this example. Productions are *rewriting rules*. The algorithm works as follows:

**RW1:** Write down the sentential symbol.

**RW2:** For each word generated so far, determine which symbols are nonterminals and which are terminals. If the word has only terminal symbols, apply RW4. If the word has nonterminal symbols, choose the productions with that symbol as the left-hand side. If there are nonterminals, remove this word from consideration. Make as many copies of the word as there are productions and replace the nonterminal with the right-hand sides.

**RW3:** Continue until there are no changes.

**RW4:** If any word matches the input word, then accept the input word.

This certainly is not very efficient, nor easy to program for that matter, but we will address that later.

We use the '$\Rightarrow$' symbol to symbolize the statement "derives." That is, read "$A \Rightarrow B$" as "$A$ derives $B$."

Now for an example. Is "1+2*3" grammatically correct? See Figure III.3

After studying the rules we can find a direct path. See Figure III.4. So, yes: "1+2*3" is grammatically correct. But I made some lucky guesses along the way. But programs can't guess; they must be programmed.

Notice that the parse given in Figure III.5 is not the only one. For example, we could have done the following:

$$E \Rightarrow T + E$$

$$\Rightarrow T + F * T$$

$$\Rightarrow T + F * F$$

| Word | Products | Comments |
|------|----------|----------|
| $E$ | $T + E, T$ | Must start with $E$ |
| $T$ | $F * T, F$ | Choose any, expand |
| $F$ | $(E), i$ | Not worth expanding; can't be right |
| $F * T$ | $i * T, (E) + T$ | ditto |
| $T + E$ | $F * T + E, F + E$ | Should get us somewhere |
| $F * T + E$ | $i * T + E, (E) * T + E$ | Not worth expanding |
| $F + E$ | $1 + E, (E) + E$ | Second one not worth it |
| $1 + E$ | $\ldots$ | This will eventually work |

**Figure III.3    Parsing $1 + 2 * 3$ by Simple Rules of Rewriting**

$$
\begin{aligned}
E \quad &\Rightarrow \quad T + E \\
&\Rightarrow \quad F + E \\
&\Rightarrow \quad 1 + E \\
&\Rightarrow \quad 1 + T \\
&\Rightarrow \quad 1 + F * T \\
&\Rightarrow \quad 1 + 2 * T \\
&\Rightarrow \quad 1 + 2 * f \\
&\Rightarrow \quad 1 + 2 * 3
\end{aligned}
$$

**Figure III.4    Nondeterministic Parse of $1 + 2 * 3$**

| | |
|---|---|
| $E \to T$ | $\{ \$\$ = \$1 + \$3 \}$ |
| $E \to T$ | $\{ \$\$ = \$1 + \$3 \}$ |
| $E \to T$ | $\{ \$\$ = \$1 \}$ |
| $T \to F * T$ | $\{ \$\$ = \$1 * \$3 \}$ |
| $T \to F / T$ | $\{ \$\$ = \$1 * \$3 \}$ |
| $T \to F$ | $\{ \$\$ = \$1 \}$ |
| $F \to$ any integer | $\{ \$\$ = \$1 \}$ |
| $F \to (E)$ | $\{ \$\$ = \$2 \}$ |

**Figure III.5    Production System for Arithmetic Expressions without Variables**

$$\Rightarrow T + F * 3$$

$$\Rightarrow T + 2 * 3$$

$$\Rightarrow F + 2 * 3$$

$$\Rightarrow 1 + 2 * 3$$

In order to keep such extra derivations out of contention, we impose a rule. We call this the leftmost canonical derivation rule (LCDR): expand the leftmost nonterminal symbol first.

### Production Systems

Production systems are useful generalization of grammars. This generalization is also a generalization of FSAs to FSMs. Production systems are an obvious outgrowth of the transformations in the section "Regular Grammar First." The transformation method we used in that section can be applied to finite state *machine* graphs, leaving us with the schema

$$N \rightarrow V\{actions\}$$

where $N$ is a nonterminal and $V$ is any number of terminals and nonterminals (because of the form of context-free productions).

Production systems are used extensively in artificial intelligence applications. They are also the design concept in parser generating tools such as *LEX* and *YACC*. The basic idea is that pattern matching must terminate successfully before any execution of program segments can occur.

Instead of a long discussion, let's work with something we know: the simple arithmetic language defined in Figure III.3. This is the hand calculator problem from the famous "Dragon Book" (Aho and Ullman 1972). The way to read

$$E \rightarrow T + E \ \{\$1 + \$3\}$$

is as follows (Figure III.5):

1. The symbols preceded by a $ (dollar sign) are semantic in intent; think of the dollar sign as a special version of the letter 's'. $$ is the semantic content of the left-hand symbol $E$, while $n$ is the semantic value of the $n$th symbol on the right-hand side.
2. The program that implements the production first applies the production $E \rightarrow T + E$. Two outcomes are possible:

a. The production may fail. Since there may be many recursions, it is possible that the production does not complete. If this is the case, then nothing further other than error processing occurs.

b. If the production succeeds, the *action* segment (after translating the symbols preceded by $'s to memory locations) is then executed.

The question should occur to you: Where is the data? The answer is, on the stack at the locations named by the $'s symbols. But the details are left to you.

## PARSER DESIGN USING PRODUCTION SYSTEMS

### ST Graphs to Production Systems

We could describe the FSA representation lexical rules by productions developed by the following rules:

- Name all the nodes of the graph. These names become the categorical symbols.
- For every arc, there is a *source node* and a *sink node*. The *terminal symbols* are the weights of the arcs.
- For every source node, for example *A*, and every sink node, for example *B*, write down a grammar rule of the form $A \to B$, where *t* is the weight of the arc. If there is no such *t*, then discard the rule.

The second requirement is to move the functions from the ST graphs. The convention that has evolved is that we write the productions as above, then enclose the *semantic actions* in braces.

In effect, we have laid the foundations for a very important theorem in formal languages: every language that is specifiable by an ST graph is specifiable by a grammar. The content of the theorem is that many apparently different ways of specifying computations are in fact the same.

### Parse Trees

Remember the idea of a derivation:

$$E \Rightarrow T + E$$

$$\Rightarrow T + F * T$$

$$\Rightarrow T + F * F$$

$$\Rightarrow T + F * 3$$

$$\Rightarrow T + 2 * 3$$

$$\Rightarrow F + 2 * 3$$

$$\Rightarrow 1 + 2 * 3$$

The derivation induces a tree structure called a *parse tree*. When thought of as an automaton, the tree is the justification for accepting any sequence of tokens. Thought of as a machine, however, the tree is not the whole story.

The productions define a *function* that does two things: it recognizes syntactically correct inputs and it can compute with the inputs. Instead of the simple derivation above, we can think about tokens for the terminals and values for the categorical symbols. For this discussion to make sense, follow the LCDR. The LCDR requires that the parse be done by always expanding the leftmost categorical symbol. We do this in a top-down manner. Such a parser is called a *predictive parser* or, if it never has to backtrack, a *recursive descent parser*. Let's try this idea on $1 + 2 * 3$. In order to make it clear that this is a production system parse, we enclose the nonterminal symbols in angular braces $\langle \cdot \rangle$. The symbolic name is on the left and the semantic values are on the right. The symbol $\perp$, read "bottom," stands for the phrase "no semantic value defined." See Figure III.6. Line 3 contains the first terminal symbol that is not an operator. The relevant production is

$$F \rightarrow \text{any integer } I\{\$\$ = atoi(\$1)\}$$

In words, it says, "the next input is an integer. Recognize it. If the integer is recognized syntactically, then perform the function *atoi* on the token $1. If the conversion works, assign the result to $$."

The parser returns to the previous level, so we have the $F$ replaced by $T$. The parser now predicts that the atom $\langle atom, + \rangle$ is next. When the parser sees the '+', it reads on.

Repeating the same reasoning, we end up with a final string of

$$\langle T, \langle int, 1 \rangle \rangle \langle atom, + \rangle \langle F, \langle int, 2 \rangle \rangle \langle atom, * \rangle \langle F, \langle int, 1 \rangle \rangle$$

| | | | |
|---|---|---|---|
| 1 | $\langle E, \perp \rangle$ | | |
| 2 | $\langle E, \perp \rangle$ | $\Rightarrow$ | $\langle T, \perp \rangle \langle atom, + \rangle \langle E, \perp \rangle$ |
| 3 | | $\Rightarrow$ | $\langle F, \perp \rangle \langle atom, + \rangle \langle E, \perp \rangle$ |
| 4 | | $\Rightarrow$ | $\langle F, \langle int, 1 \rangle \rangle \langle atom, + \rangle \langle E, \perp \rangle$ |
| 5 | | $\Rightarrow$ | $\langle T, \langle int, 1 \rangle \rangle \langle atom, + \rangle \langle E, \perp \rangle$ |
| 6 | | $\Rightarrow$ | $\langle T, \langle int, 1 \rangle \rangle \langle atom, + \rangle \langle T, \perp \rangle$ |
| 7 | | $\Rightarrow$ | $\langle T, \langle int, 1 \rangle \rangle \langle atom, + \rangle \langle F, \perp \rangle \langle atom, * \rangle \langle E, \perp \rangle$ |
| 8 | | $\Rightarrow$ | .... |

**Figure III.6  Derivation Using Production System**

and we start returning. We return to a level that has the '∗' in it. The correct computation takes the 2 and 3 and multiplies them. So we get

$$\langle T, \langle int, 1 \rangle \rangle \langle atom, + \rangle \langle T, \langle int, 6 \rangle \rangle$$

We can now compute the last operation and get 7. So

$$\langle E, \langle int, 7 \rangle \rangle$$

is the final value.

# PARSER DESIGN II

Work on the theory of computability shows that many different forms of algorithm presentation are equivalent in the sense that one form can be converted to another. This is a very powerful idea and we will approach this from a categorical point of view.

## Theoretical Underpinnings

The conversion (compilation if you will) we want to make is between the production system and a typical programming language. The background is based on work on the Chomsky Hierarchy. The Chomsky Hierarchy is a sub-theory of the theory of computation that indicates what abstract computing models are equivalent in the sense that programs in one model produce the same outputs as a particular algorithm specified in other models. In this case, we are interested in two models: production systems and recursive functions. All current programming languages are examples of recursive function systems. Therefore, we will make the task one of going from the formal definition of production systems to some sort of programming language.

We know that there are four parameters in the grammar: $N$, $T$, $G$, $S$. Consider now trying to write a program. What is being specified by $N$, $T$, $G$, and $S$? The key to Subtask 1 is to understand that the categorical symbols (symbols in $N$) are really the names for programs and that the programs implement the rules in $G$.

# CASE 16. CONVERTING THE *S* PRODUCTION TO CODE

Use pseudo-code to convert the $S$ production of the grammar to a program.

## TESTING THE PARSER: CONTEXT-FREE LANGUAGE GENERATION

The possibility exists of writing a program to develop test inputs for the parser. The design of such a program indicates how we can think about these tests.

The term *language* in our context means a set of *words* that are comprised of only terminal symbols and those words must be ordered in the prescribed fashion. However, there are many intermediate forms derived during a parse. Since we're using a recursive descent parser with a programming discipline of writing a recursive function for each categorical symbol, the appearance of a categorical symbol within the words indicates where programs are called. The key to developing tests is that the parser must work from left to right.

For formality's sake, we can define the *length of a sentence* as the number of (terminal or categorical) symbols in the word. For example, the statement $(1 + 2)$ has length 5. The key to testing, then, is to think of various ways you can generate words that are only comprised of terminal symbols in such a way that all the possible categorical symbols are also involved.

## CASE 17. STRUCTURAL INDUCTION OF WORDS

What is the shortest length terminal word? Show its derivation. What is the next shortest? Show its/their derivation.

Demonstrate a terminal word of length $n = 1, ..., 5$ or show that it is impossible to have a word of that particular length.

Outline how you could rewrite the parsing algorithm to a program that can generate the legal strings of any length. Generate five unique test strings.

We could use the input grammar specified in the specifications to read in the SOL statements. However, one reason to choose a language based on Cambridge Polish is that we can use a very simple input grammar and push many issues into the semantic phases. Close inspection of Cambridge Polish shows that it is just the language of balanced brackets with simple terminal words (names, numbers, strings, etc.) sprinkled throughout.

## CASE 18. DEVELOPING THE PARSING GRAMMAR

One grammar that recognizes balanced brackets is shown below. Let $S$ be the *sentential symbol*.

$$S \rightarrow []$$
$$S \rightarrow [S]$$
$$S \rightarrow SS$$
$$S \rightarrow Atom$$
$$Atom \rightarrow integer$$
$$Atom \rightarrow floating$$
$$Atom \rightarrow string$$

**Exercise.** Convince yourself that the grammar actually does what is purported.

**Exercise.** Modify the grammar to accept the lexical entities outlined in the SOL specification.

**Assignment.** Take the grammar and produce a production system that parses incoming tokens and builds a tree that represents the actual *parse tree*.

# Milestone IV

## TYPE CHECKING

**Reference Materials**

Terrence W. Pratt and Marvin V. Zelkowitz. *Programming Languages: Design and Implementation.* 4th Ed. Upper Saddle River, NJ: Prentice Hall. 1995. Section 4.2; Chapters 5 and 6.

Kenneth C. Louden. *Programming Languages: Principles and Practice.* 2d Ed. Belmont, CA: Thompson, Brooks/Cole. 2003. Chapters 6 and 9.

## OBJECTIVES

**Objective 1:** To produce a static type checker for the language

**Objective 2:** To extend the symbol table to include type information

**Objective 3:** To give you experience using semantic formalisms in designing the type checker

**Objective 4:** To gain experience in modifying and enhancing your programs

**Objective 5:** To develop personal project management skills

## PROFESSIONAL METHODS AND VALUES

Type checking is the first step in semantic processing. We can again use a production system schema, but this time we will pattern match on node content. Designing the type checker requires that we enumerate all the possible legitimate patterns we might see; other patterns encountered constitute type errors.

## ASSIGNMENT

If we had only constants and primitive operations in SOL, then we could easily generate the Forth input by traversing the tree in postorder. However, this would not be very usable because we would not have variables or user-defined functions. Once we introduce variables and functions, we introduce the need to know what types of values can be stored in variables and what types of values can be used with functions. These points introduce the need for types and type checking. If this were a language like C, then we would now have to do several things:

- Modify the scanner to accept new specially spelled lexemes
- Modify the parser to accept new syntax and construct new trees
- Modify the symbol table to develop new information

The first two are not necessary in SOL because the syntax of the language is so simple, but we will have to elaborate the symbol table information.

The type checker produces (possibly) new trees based on decisions made due to overloading of primitive operators and polymorphism in user-defined functions. This decision logically goes here since the type checker has the correct information in hand at the time of the decision.

## PERFORMANCE OBJECTIVES

1. To develop a formal definition of the type checker from the intended (naïve) semantics for SOL
2. To test the resulting program for correctness based on the formal definition

## MILESTONE REPORT

The milestone report should concentrate on the development of the program. It must include a copy of your formal design documents.

## WHAT IS A TYPE?

The importance of types was not appreciated in assembler or early programming languages. Starting in the late 1960s and early 1970s, programming languages included formal mechanisms for inventing user-defined types. ML was the first language to emphasize types and its initial proposal is attributed to Robin Milner. The extensive use of types is now a hallmark of functional languages, whereas the C-based languages are somewhere in-between.

The concept of types originated with Bertrand Russell (1908) as a way out of his famous paradox in set theory. Various uses of type theory have evolved since then, with its use in logic being the most like that of programming languages. A type $T$ is a triple $(T, F, R)$ where $T$ is a set of objects, $F$ is a set of functions defined on $T$, and $R$ is a set of relations defined on $T$. In what is known as an "abuse of language," the name of the type $T$ is the same as the name of the set in the definition. This abuse is often confusing to people first learning programming.

We shall not be interested in the theory of how such types are developed and justified. However, since we are developing a functional language, we must understand how to tell whether or not an expression is well formed with respect to types. For example, we would normally say that an expression such as 5 + "abcd" is incorrectly typed primarily because we cannot conceive of a sensible answer to the expression. When all is said and done, that is the purpose of types: to guarantee sensible outcomes to expressions.

Types come naturally from the problem that the program is supposed to solve. Types in programming are conceptually the same as *units* in science and engineering. A recent mistake by NASA illustrates the importance of each of these ideas.

> The *Mars Climate Orbiter* was destroyed when a navigation error caused the spacecraft to miss its intended 140–150 km altitude above Mars during orbit insertion, instead entering the Martian atmosphere at about 57 km. The spacecraft would have been destroyed by atmospheric stresses and friction at this low altitude. A review board found that some data was calculated on the ground in English units (pound-seconds) and reported that way to the navigation team, who were expecting the data in metric units (newton-seconds). *Wikipedia.*

In this case, *units* play the same role as *types*.

## REASONING ABOUT TYPES

As data structures become more complex, the type system becomes more important. Early programming languages had little or no support for types. Early Fortran and C compilers assumed that undeclared variables were integers and that undeclared functions returned integers. In fact, Fortran had a complicated system of default types based on the first letter of the name.

The programming language ML was originally developed in about 1973 at Edinburgh University for the research community as part of a theorem-proving package called LCF (Logic of Computable Functions). Syntactically, ML looks very much like an earlier language, Algol. The Algol language may have had more impact on programming languages than perhaps any other. Two versions of Algol came out close to one another, in 1958 and 1960. A third came out in 1968 that would be the last, but even Algol 68 (as it was known) had a lasting impact on Unix: the Bourne shell. ML demonstrated that a strongly typed programming language was practical and efficient.

## CASE 19. HISTORICALLY IMPORTANT LANGUAGES

Using the resources available to you, develop a short history of Algol and ML. Also, develop a list of areas these languages impacted and continue to impact today.

## DEVELOPING TYPE RULES

Like SOL, ML allows the programmer to define types that are combinations of primitive types, user-defined types, and type constructors. ML does not require the user to type declare user names because the compiler is able to infer the correct type—a good reason to use languages such as ML.

ML is a strongly typed language, meaning that the compiler type checks at *compile* time and can reliably infer correct types of undeclared names. This is decidedly different from languages such as Fortran and C that require the user to declare types for all names. Eventually, SOL will be strongly typed, but our project does not require us to develop the algorithm to produce the correct inferences. For this milestone, we will consider type coercion but not polymorphism. (Ah, but coercion is a type of polymorphism.)

ML supports polymorphic functions and data types, meaning that you need only define a type or function once to use it for many different data or argument types, avoiding needless duplication of code. This gain in production is from the use of *type variables*; C++ templates are a poor substitute for type variables. C has no such capability: each such need requires that you write the code and that you name the code uniquely. Java's polymorphism would require many methods, each taking different data types and performing similar operations. In other words, C-type polymorphism (C, C++, Java, among others) provides no help in development, requiring the programmer to design, develop, and manage many different aspects that the compiler should be doing.

ML also provides support for abstract data types, allowing the programmer to create new data types as well as constructs that restrict access to objects of a certain type, so that all access must take place through a fixed set of operations. ML also allows for type variables, variables that take on types as their values. ML achieves this through the development of *modules* and *functors*.

The original development of ML was spurred by the need to have a complete type system that could infer correctly undeclared names. This type system is completely checked at compile time, guaranteeing that no type errors exist in the resultant code.

## CASE 20. ML TYPE SYSTEM

Research the ML type system using resources at your disposal. Report on three aspects of the system:

1. How does the compiler infer the correct type of an undeclared variable?
2. What are the strengths of the ML type system?
3. What are the weaknesses of the ML type system?

## DESIGN ISSUES

The first version of SOL must imitate the type checking used by C. Since C is so common, we leave it up to the designers to find the detailed rules. Detailed information on designing rule-based systems is given in Chapter 8.

The metalanguage for type checking is that of logic. Type checking is about *inference,* which is inherently rule-based. Modern expositions in logic use a *tree* presentation. The presentation of the rules follows a simple pattern. You know that logic revolves around the *if-then* schema: *if A then B.* This would be displayed as

$$\frac{A}{B} MP$$

In the context of type and type rules, an example would be for plus.

$$\frac{integer \quad integer}{integer} + integer$$

## DEVELOPING TYPE-CHECKING RULES

*Every* possible combination of types for any given operator must be considered. Every legitimate combination of types must have a rule and every illegitmate combination must have a rule describing the error action or correction.

A simple approach to analyzing and developing these rules is to develop a matrix of types. We have a very simple type system to consider initially: `bool`, `integer`, `float`, `string`, and `file`. Every unary operator must be considered with each type; every binary operator must be considered with each *pair* of operators.

## TYPE CONVERSION MATRIX

Develop a matrix of types as described above. In each element of the matrix, write the correct Forth conversion statement. For incompatible elements, write a Forth `throw` statement.

In SOL, the standard operators are overloaded, meaning they automatically convert differing types into a consistent—and defined—type, which is then operated on. This feature follows mathematical practice but has long been debated in programming languages. Pascal did not support automatic type conversion, called mixed-mode arithmetic at the time. SOL does support mixed-mode expressions (just like C); therefore, the design must include type checking and automatic conversion. Because of overloading, there are implicit rules for conversions. The design issue is to make these implicit rules explicit.

Standard typography in textbooks as described above does not have a standard symbology for this automatic conversion. Therefore, we'll just invent one. It seems appropriate to simply use functional notation. All the type names are lower case, so upper-case type names will be the coercion function. This doesn't solve the whole problem, since it must still be the case that some conversions are not allowed.

As an example,

$$\frac{\text{FLOAT}(\text{integer}) \quad f\text{loat}}{f\text{loat}} + \text{integer-float conversion}$$

Such a rule is interpreted as follows. An example to work from is [+ 1 3.2]. The implicit rule requires that the integer be converted to a float. We note that by having FLOAT(integer).

An example of something that should not be allowed would be

$$\frac{\text{FLOAT}(\text{integer}) \quad \text{float}}{\text{float}} + \text{integer-float conversion}$$

## SPECIFICATION FOR THE WHOLE TYPE CHECKER

Because the SOL language is a functional language, virtually every construct returns a value: the if statement, for example. Therefore, the entire expression must be checked, starting with the let and the type statements. Did you notice that there is no return statement? Type checking SOL is far more complicated than checking simple arithmetic statements because the if statement can only return *one* type. Therefore, [if *condition* 1 2.0] is improperly typed.

In order to design the whole type checker, you will have to resort to a design technique known as structural induction. Structural induction requires that we develop a tree of all the subexpressions that make up an expression. For example, what are all possible expressions that can start with a let? Here is a partial analysis.

1. The general form of the `let` statement is

   [let  [ *definiendum definiens* ]   ... ]

2. The definiendum can be one of two forms: a simple variable *name* or a *functional expression*.

   [let  [  [a expresssion] ... ]  ]
   [let  [  [ [a *arguments*] expresssion] ... ]  ]

There could be no, one, or many arguments, of course.

The point here is that you must develop a recursive function that can traverse any proper input: this is the reason to develop the printer program in the parser milestone. This printer program is the scheme for all the other milestone programs.

## CASE 22. FORM OF INPUT

It is simple to read the input if you do not try to put any semantic meaning onto the result. Such processing will not work for type checking because the type checking rules are driven by the form and content of the structures. In order to design the type checker, you must first describe the possible structures through structural induction.

Using the grammar from Milestone III, and using substitution, develop a design of the type checker, making each "nonterminal" a recursive function.

## THE SEARCH ALGORITHM

The algorithm for the type checker is more complicated than the parser. The parser is not comparing the structure to another structure, whereas the type checker is checking an expression against a rule. For example, consider the question of whether or not [+ 1 2.5] is correctly typed.

In order to discuss the problem, we first have to realize that there are different addition routines, one for each possible combination of allowed arguments. Say there are three allowed arguments: `bool`, `integer`, and `float`. This means there are nine possible combinations of arguments for '+'.

## CASE 23. SHOULD BOOLS BE ALLOWED?

While we may want to consider `bools` in the arithmetic expressions, should they be allowed? Mathematical convention would argue for inclusion.

| Forth Word Name | Return Type | Argument 1 | Argument 2 |
|---|---|---|---|
| +_int_int | int | int | int |
| +_int_float | float | int | float |
| +_float _int | float | float | float |
| +_float _float | float | float | float |

**Figure IV.1  Design Matrix for Addition**

Regardless of the outcome of the above case, there is more than one possible legitimate argument type situation. Suppose we decide that bools should not be used; then there are still four separate possibilities, given in Figure IV.1. We can take advantage of Forth's flexibility in word names by defining a unique operator for each pair. For example:

```
: +_int_float ( arg n -- ) ( float f -- f )
    s>d d>f f+
;
```

This is simple enough for each element in the type conversion matrix.

## UNIFICATION AND TYPE CHECKING ALGORITHM

Let's consider how the full type checking algorithm should proceed.

1. The type checking algorithm is entered with a tree with form

$$[ a_0\ a_1\ \ldots\ a_n ]$$

   Based on the Cambridge Polish scheme, $a_0$ must be a function and the $a_i$, $i = 1, \ldots, n$ can be either atoms or expressions.
2. Because $a_0$ must be predefined, it should be in the symbol table.
3. Suppose, for example, that $a_0$ is +. The symbol table entry must contain the list of possible types as outlined in Figure IV.1.
4. We need to search the list of possible types to find the (first) one that matches the arguments $a_1, \ldots, a_n$. In the example at hand, $n = 2$.
5. Recursively determine the type of $a_1, \ldots, a_n$. Match these types to the list of possible types.
6. When we find a match, we know both the return type and the operator that should be applied.
7. Modify the tree to replace $a_0$ with the correct operation.
8. Return the type.

This algorithm is a simplifed form of the *unification algorithm*, the fundamental evaluation algorithm used in Prolog. The above algorithm is much simpler because we do not have type variables as do ML and the other functional languages.

## CASE 24. TYPE CHECKING AND THE SYMBOL TABLE

The type checker will rely on the symbol table to produce the list of possible functions that can be substituted for $a_0$. This case and the next should be completed together, because the two solutions must be able to work together.

For the symbol table, develop a data structure for the operators. Keep in mind that when we introduce user-defined functions, they will have the same capabilities as the primitive operators.

Along with the data structure, you must develop a process to load the symbol table initially. For hints, consider how the SOL type definition process works by consulting the specification.

## CASE 25. DEVELOPING THE TYPE CHECKING ALGORITHM

On approach to developing the unification algorithm, you need really only consider a few cases. One specific case is

```
[ +   1   2 ]
[ +   1   [ *   3   4 ] ]
[ +   [ *   3   4 ] 5 ]
```

Based on these cases, develop the search algorithm and complete the design of the symbol table.

# Milestone V

## ELEMENTARY COMPILING: CONSTANTS ONLY

### Reference Materials

Terrence W. Pratt and Marvin V. Zelkowitz. *Programming Languages: Design and Implementation.* 4th Ed. Upper Saddle River, NJ: Prentice Hall. 1995. Chapters 5 and 8.

Kenneth C. Louden. *Programming Languages: Principles and Practice.* 2d Ed. Belmont, CA: Thompson, Brooks/Cole. 2003. Chapter 8.

## OBJECTIVES

**Objective 1:** To produce output suitable for use by Forth

**Objective 2:** To compile constant only operations (*No program should contain variables for this milestone.*)

**Objective 3:** To give you experience using semantic formalisms in designing the code generator

**Objective 4:** To test the actual use of the parse tree

**Objective 5:** To gain experience in modifying and enhancing your programs

**Objective 6:** To develop personal project management skills

## PROFESSIONAL METHODS AND VALUES

The principal formal concept in this milestone is the use of *structural induction*, discussed in Chapter 6. As with the parser and type checker, the general principle is the production system schema: pattern match to guarantee consistency, then application of a transformation.

# ASSIGNMENT

We are now ready to produce a minimal compiler. We have developed the parse tree and we can print that tree; the outline of the printer routine is the basis for this milestone. We now finish the first development cycle by adding a code generator.

The type checker has constructed a tree that is type correct. In the previous milestone, the type checker chose the correct polymorphic routine name.

# PERFORMANCE OBJECTIVES

1.  Develop a formal definition of the code generation algorithm from the intended (naïve) semantics for SOL based on the previously developed parser.
2.  Test the resulting program for correctness based on the definition.

For this milestone, you generate Forth code and run that code to demonstrate the correctness of the result. Your tests must show that the operations correctly implemented based on their customary definitions in boolean, integer, floating point, string and file operations. A basic approach for testing these primitive operators is to choose *simple* values that you can calculate easily. It is not a valid test case if you can't predict the result ahead of time. Therefore, you should keep the following in mind: keep the test statements short.

For the most part, this milestone is a straightforward binding exercise: choosing which word to bind to the SOL operator of a given type. This is primarily an exercise in loading the symbol table with correct information.

The details of the primitive operators are given in Figures V.1, to V.5. The special values are listed at the bottom of Figure V.5.

| Operator Name | Operator Symbol | Semantics |
|---|---|---|
| | *Primitive Data* | |
| | *When no semantics are given, C is intended* | |
| Boolean | | |
| And | A & B | Same as && in C |
| Or | A \|\| B | |
| Not | A | |
| Implies | A => B | False when A is true and B is false |
| Biconditional | A <=> B Logical equals | |

**Figure V.1 Booleans**

| | *Integers* | |
|---|---|---|
| Plus | $A + B$ | |
| Minus | $A - B$ | |
| Unary Minus | $-A$ | |
| Times | $A * B$ | |
| Divide | $A/B$ | |
| Remainder | $A\% B$ | |
| Power | $A \wedge B$ | Integer A raised to an integer power B |
| Less | $A <$ | |
| Equal | $A = B$ | Same as C's $===$ |
| Greater | $A > B$ | |
| Less than or equal to | $A \leq B$ | |
| Greater than or equal to | $A \geq B$ | |
| Not equal | $A = B$ | |

**Figure V.2    Integers**

| | *Floating Point* | |
|---|---|---|
| Plus | $A + B$ | |
| Minus | $A - B$ | |
| Unary Minus | $-A$ | |
| Times | $A * B$ | |
| Divide | $A/B$ | |
| Remainder | $A\% B$ | |
| Power | $A \wedge B$ | A float, B either float or integer |
| Less | $A <$ | |
| Equal | $A = B$ | Same as C's $===$ |
| Greater | $A > B$ | |
| Less than or equal to | $A \leq B$ | |
| Greater than or equal to | $A \geq B$ | |
| Not equal | $A = B$ Sine | $\sin(A)$ |
| Cosine | $\cos(A)$ | |
| Tangent | $\tan(A)$ | |
| Exponential | $\exp(A)$ | $e^A$ |

**Figure V.3    Floats**

## STRING IMPLEMENTATION

One nontrivial issue is the full development of string support. The formal concept here is CRUD: creating, reading, updating, and destroying strings. Creating and destroying string constants introduces two storage

|  | *Strings* |  |
|---|---|---|
| Concatenation | $A + B$ |  |
| strcat in C |  |  |
| Insert | insert($A, B, C$) |  |
| Insert character $B$ at location |  |  |
| $C$ in string $A$ |  |  |
| Character of | charof($A, B$) | Return the $B$th character of string $A$ |
| Less | $A <$ |  |
| Equal | $A = B$ | Same as C's === |
| Greater | $A > B$ |  |
| Less than or equal to | $A \leq B$ |  |
| Greater than or equal to | $A \geq B$ |  |
| Not equal | $A = B$ |  |

**Figure V.4   Strings**

|  | *Files* |  |
|---|---|---|
| Open | open($A$) | $A$ is a path expression as a string |
|  |  | Returns a file  value |
| Close | close($A$) | Closes open file  $A$ |
| Test end of file | endoffile ($A$) | returns true is file  $A$ is at end of file |
| Read boolean value | readbool($A$) | File $A$ is read for boolean |
| Write a boolean value | writebool($A, B$) | Write boolean value $B$ to file  $A$ |
| Read integer value | readint($A$) | File $A$ is read for integer |
| Write a integer value | writebool($A, B$) | Write integer value $B$ to file  $A$ |
| Read float  value | readfloat ($A$) | File $A$ is read for float |
| Write a float  value | writefloat ($A, B$) | Write float  value $B$ to file  $A$ |
| Read string value | readstring($A$) | File $A$ is read for string |
| Write a string value | writestring($A, B$) | Write string value $B$ to file  $A$ |
|  | *Special Values* |  |
| Pi | Pi | Closest representable value of $\pi = 3.14\ldots$ |
| Standard in | stdin |  |
| standard out | stdout |  |

**Figure V.5   Files and Special Constants**

management issues: (1) obtaining and releasing memory and (2) knowing when it is safe to destroy the string.

## IF AND WHILE

This is the appropriate place to implement if and while statements. Admittedly, the testing is fairly trivial, except for the while; complete testing of the while must wait.

## MILESTONE REPORT

The milestone report should concentrate on the development of the program. It must include any design decisions that you make that are not also made in class; all such decisions should be documented by explaining why the decision was made.

# Milestone VI

---

# SCOPING AND LOCAL VARIABLES

---

**Reference Material**

Terrence W. Pratt and Marvin V. Zelkowitz. *Programming Languages: Design and Implementation*. 4th Ed. Upper Saddle River, NJ: Prentice Hall. 1995. Chapters 6 and 7; Section 9.2.

Kenneth C. Louden. *Programming Languages: Principles and Practice*. 2d Ed. Belmont, CA: Thompson, Brooks/Cole. 2003. Chapter 8.

---

## VARIABLES AND LOCAL VALUES

### Objectives

**Objective 1:** To produce output suitable for use by Forth

**Objective 2:** To introduce user-defined names

**Objective 3:** To give you experience using semantic formalisms in designing the code generator

**Objective 4:** To gain experience in modifying and enhancing your programs

**Objective 8:** To develop personal project management skills

### Professional Methods and Values

Local names and values are a binding issue: Under what circumstances does the evaluation of a name result in a correct value? The design and implementation of local variables revolves around understanding how the execution sequence of a program is influenced by calls and returns.

## Assignment

We now have a compiler that can take in any arithmetic statement and can generate code for the two obvious commands: *if* and *while*. We now want to add commands that allow the programmer to have meaningful names. There are some issues:

1. Is the variable's scope *global* or *local*?
2. Is the variable's value *persistent* or *transient*?
3. How to reference the value of a local variable in the local activation.
4. How to reference the value of a local variable in a nonlocal activation.
5. How to reference the value of a global variable.
6. How to assign values to either global or local variables.

## Performance Objectives

1. Develop a formal definition of the code generation algorithm from the intended (naïve) semantics for SOL based on the previously developed integer code generator.
2. Test the resulting program for correctness based on the formal definition.

## Milestone Report

The milestone report should concentrate on the development of the program. It must include a copy of your formal design documents.

## SCOPING AND SOL

The concept of *scope* is intimately connected to the concept of *visibility of names* and with storage management. Storage management issues are discussed in detail in Chapter 10. Scope rules determine how long a name is bound to a value.

Specifically in SOL, there are several types of scoping environments inherited from most block-structured languages such as C. For this milestone, we have the following scope protocols:

1. File scope
2. Function header scope
3. Function body scope

The function header scope is discussed separately because it is a composite of file scope and body scope.

| Scope Protocol | Variables | Functions |
|---|---|---|
| File Scope | | |
| Header Scope | | |
| Body Scope | | |

**Figure VI.1    Analysis of Scope-Storage Interactions**

## CASE 26. BINDING TIMES AND SCOPE

Consideration of scope comes from considering the binding time of a name with its definition. In the simple programming language context, a name can only have one active binding at a time: otherwise, there is ambiguity. We should note that multiple inheritance in object-oriented languages such as C++ introduces ambiguity. Therefore, binding time requirements lead to the need for scoping. What properties should *scope* have?

The original idea of scope and binding time came from logic. The introduction of *quantifiers* by Pierce, Frege, and others in the 1880s introduced the problem of name-definition ambiguity. The basic issue in logic is the nesting and duration of definitions. Programming adds another dimension: storage. Whenever a name is defined, there is some consequence for memory management. Although individual routines are accounted for at compile time, variables have other issues, such as access and duration of a value. All nonpersistent, local storage must be reclaimed when the execution leaves the scope.

## CASE 27. SCOPES: FILLING IN THE DETAILS

For each of the scope protocols listed above, determine the consequences for memory management. Do this by displaying a table (as shown in Figure VI.1) that lists the rules for each possibility.

## COMPILING ISSUES FOR SOL

### Symbol Table Issues

Introduction of local variables and scope alters the way we think about the symbol table. Up to this point, we have treated the symbol table as a repository for system-defined names and symbols. It was used primarily in its lookup capacity. That no longer is possible.

To understand the issue, consider again the issues of scope; more to the point, nested scopes. A simple C program segment is shown in Figure VI.2: the two i definitions are completely legitimate because the left brace opens a new scope. The symbol table must account somehow for this nesting.

```
{
    int i = 5;
    int j = 7;
    ...
    {
        inti = 12;
        printf("%d", i + j);
    }
}
```

**Figure VI.2  Illustration of Scoping Issues**

## CASE 28. HANDLING NESTED SCOPES

Two different approaches come to mind on how to handle the nesting of scopes. One would be to have a separate symbol table for each scope. The second approach would be to make the symbol table more complicated by keeping track of the scopes through some sort of list mechanism. One such mechanism would be to assign a number to each different scope and use the number to disambiguate the definitions. This approach keeps a stack of numbers. When a scope opens, the next number is placed on the stack; when the scope closes the current number is popped and discarded. Every definition takes as its scope the number at the top of the stack.

Use each of these approaches in a design and determine the running time ("Big O") and the space requirements for each. From this, make a decision on which approach you will use in this milestone.

### Space Management Issues

SOL has an important problem for local variable compilation: the local variables "live" on the stack. However, Forth has several different stacks: the *argument stack* for integers and integer-like values (such as pointers) and a *floating point* stack. There are other stacks that are handy to know about, such as the return stack and the control stack. For this milestone the argument and floating stacks are the focus. To illustrate an issue, the routine with header

```
float foo(int a, float b)
```

must have its a argument on the argument stack and the b argument on the floating point stack; the return value must be pushed onto the floating point stack.

This problem is made worse by the fact that local variables are also on the stack. In fact, the local variables are defined early on in the computation, so the compiler must be able to determine where a given value is and how to move that value to the top of the stack, as well as assign the top of the stack to a variable.

In effect, the compiler must simulate the running of the program because the dynamic program will be moving values around on the stack in accordance with the instructions laid down by the compiler.

## CASE 29. GETTING TO THE DETAILS: SCOPE AND FORTH

Develop a *simple* SOL program that is comprised only of integer arguments and integer variables; three of each should be sufficient. Now write several arithmetic operations that compute values and store them in local variables. Write the equivalent Gforth program. Use this experience to develop the "simulator" needed to compile code.

The basic issue facing the compiler in this milestone, then, is the simulation of the stack. In order to simulate the stack, you must have a list of rules of the effect of each operation on the stack. This is actually a straightforward exercise: develop a table that lists the operations on the rows and stacks information across the columns.

## CASE 30. STACK EFFECTS

Develop a *stack effect table* using the following outline as a starting point:

| Operation | Input | | Output | |
|---|---|---|---|---|
| | Argument | Floating | Argument | Floating |

Your design should designate operations for *all* defined operations, including those you may have introduced to simplify type checking.

Once the table is complete, study the resulting pattern of operations. There are relatively few unique patterns. A simple approach to the problem of looking up what pattern to use is to extend the information in the symbol table to include the stack information.

# Milestone VII

## USER-DEFINED FUNCTIONS

---

**Reference Materials**

Terrence W. Pratt and Marvin V. Zelkowitz. *Programming Languages: Design and Implementation.* 4th Ed. Upper Saddle River, NJ: Prentice Hall. 1995. Chapter 9.
Kenneth C. Louden. *Programming Languages: Principles and Practice.* 2d Ed. Belmont, CA: Thompson, Brooks/Cole. 2003. Chapters 7 and 8.

---

### OBJECTIVES

**Objective 1:** To produce code generation for nonrecursive and recursive functions

**Objective 2:** To extend the symbol table to include function header information

**Objective 3:** To give you experience using semantic formalisms describing lambda expressions

**Objective 4:** To gain experience in modifying and enhancing your programs

**Objective 5:** To develop personal project management skills

The principles invoked in this milestone are the relationship of the λ-calculus and program evaluation.

### ASSIGNMENT

We now have a complete compiler, but not a very usable one because we do not have functions. You must now include the headers for functions, arguments, and return types. We use the specification for the syntax and use your understanding of programming to produce the semantics.

## PERFORMANCE OBJECTIVES

1. Develop a formal definition of the intended (naïve) semantics for SOL based on the previously developed assembler and your knowledge of programming.
2. Test the resulting program for correctness based on the formal definition.

## MILESTONE REPORT

The milestone report should concentrate on the development of the program. It must include a copy of your formal design documents.

## BINDING THE FUNCTION NAME TO THE FUNCTION

Chapter 9 describes the fundamental theoretical understanding of functions and function evaluation: the λ-calculus. Some languages, such as ML and OCAML, use the descriptive term *anonymous function* to describe the role of the λ-calculus: to describe function bodies. But how does one bind the λ-expression to a name?

This actually brings up a deeper issue: How do we think about binding any name to its definition? The basic approach is not unlike normal *assignment* in a program. For example, in the C statement int a = 5; we think of the location denoted by a as naming the location of integer 5. Does this work for functions? For example, how can one rename a function? How about

```
double(*newsin)(double) = sin;
```

If we then write newsin(M_PI/4) do we compute sin π/4? Of course.

We're almost there. Suppose we don't want the value of newsin to be changeable. Again in C, the way to do that is to use const in the right place.

```
double(* const newsin)(double) = sin;
```

The purpose of this exercise is to demonstrate function names are nothing special, but before we finish this investigation, what is happening in terms of the λ-calculus? What is sin in terms of λ-calculus? Without investigating algorithms to compute the sine function, we can at least speculate that the name sin is a λ-expression of one argument. In other words, the assignment of sin to newsin is simply the assignment of the λ-expression.

The problem is type. Chapter 9 does not explore the issue of types, leaving it intuitive. The λ-calculus was introduced to explore issues in the natural numbers and integers. The development of the expanded theory came later but, for our purposes, can be seen in such languages as Lisp, ML, and OCAML. Types are bound to names using the colon ':'. Thus,

$$\lambda x : \text{ integer. } \textbf{body} : \text{ integer}$$

directs that the argument x is an integer and that the result also be an integer.

## RECURSIVE AND NONRECURSIVE FUNCTIONS

Early in the history of computer science, special status was accorded to recursive functions. Generally, recursive functions were treated as something to avoid, and often were required to have special syntax. C went a long way to change that since it has no "adornments" on the functions' header. We take this approach in SOL: all functions are by nature recursive.

This approach works fine for the naïve approach, but is not suitable to the theoretical underpinnings. In the theoretical literature, the following issue raises its head:

$$let \; fibonacci(n) = \lambda n.$$
$$if \; n < 2 \; then \; 1 \; elsefibonacci(n-1) + fibonacci(n-2)$$

Notice that there are three instances of the word *fibonacci,* one on the left and two on the right. Question: How do we know that the one on the left is invoked on the right? Answer: Nothing stops us from the following:

$$let \; foobar(n) = \lambda n.$$
$$if \; n < 2 \; then \; 1 \; elsefibonacci(n-1) + fibonacci(n-2)$$

in which case *foobar*(2) is undefined. For this reason, languages based on the λ-calculus have a second `let` statement. This second statement is written `letrec`; the `letrec` guarantees that the recursive name is properly bound. In keeping with this convention, extend SOL to include `letrec`.

# IMPLEMENTATION ISSUES

Functions add three issues to the implementation: (1) passing of arguments, (2) the return of computed values, and (3) storage management. It should be clear that the stack is the communications medium.

## Parameter Passing and Value Returning

Because Gforth is a stack-based system, the development of protocols for parameters and returns is forced on us.

# CASE 31. PASSING PARAMETERS IN SOL

The protocol for passing arguments is not particularly intricate for a simple implementation such as ours. Because of the ability to type arguments, there should be no type mistakes. You will have to develop rules for arguments that must reside on different stacks, such as the *argument* and *float* stacks.

Write down the rules for passing arguments of the primitive types. One issue to address is the question of what the stack state(s) should be and when the final states are assumed.

Returning values is more complicated than passing because the *calling routine* must retrieve those values and dispose of them properly, leaving the *stacks* in their correct state. Potentially, there could be many return values for a given call. However, if you think about it carefully, you can see that returning values is not completely unlike passing arguments.

# CASE 32. RETURN VALUES

Modify the protocol for passing arguments to include the return protocol. For this initial compiler, assume that only one return value is possible.

What will you do if there is *no* return value?

## Storage Management Issues

Milestone VII provided the basic mechanisms for stack management *within* the body of a function. However, there is still a question of what should happen to the parameters themselves? The answer is clear: the basic concept of a function in our constructive world is that we replace the function and its parameters with the result. Therefore, the stack must be returned to the state it was in *before* the first parameters were pushed on *and then* the return value is placed on the correct stack.

The last issue in this regard is the assignment of the result once control has returned to the *calling routine.* There are several possibilities:

1. [assign a [foo x]]. In this case, assign naturally accounts for the extra stack location.
2. [assign a [foo x]] and [foo x] returns no value. In this case, we have a type error.
3. [foo x] returns a value but it is not assigned to any name. In this case the value must be popped; the *called* routine does not have any information about the use of the values.

# Milestone VIII

## USER-DEFINED COMPLEX DATA

### Reference Materials

Terrence W. Pratt and Marvin V. Zelkowitz. *Programming Languages: Design and Implementation*. 4th Ed. Upper Saddle River, NJ: Prentice Hall. 1995. Chapter 9.
Kenneth C. Louden. *Programming Languages: Principles and Practice*. 2d Ed. Belmont, CA: Thompson, Brooks/Cole. 2003. Chapters 7 and 8.

## OBJECTIVES

**Objective 1:** To produce code generation for complex data types: arrays and structures

**Objective 2:** To extend the symbol table to include type information for complex data organizations

**Objective 3:** To give you experience using semantic formalisms describing data referencing operations

**Objective 4:** To gain experience in modifying and enhancing your programs

**Objective 5:** To develop personal project management skills

## PROFESSIONAL METHODS AND VALUES

The methods and values of this milestone harken back to the days when assembler languages were all we had. In assembler language, all processing is carried out by the programmer. Designing, implementing, and manipulating complex data structures at the assembler level are fraught with issues requiring complete concentration. Users of modern programming languages have extensive libraries for manipulating standard data structures (such as C's `stdlib`). However, such libraries cannot fulfill all contingencies; therefore, compilers must support user-defined complex structures.

# ASSIGNMENT

We now have a complete compiler, including functions, but we do not have facilities for user-defined complex data. We use the specification for the syntax and now use your understanding of programming to produce the semantics, primarily the semantics of pointers and pointer manipulation.

# PERFORMANCE OBJECTIVES

1. Develop a formal definition of the intended (naíve) semantics for SOL based on the previously developed assembler and your knowledge of programming.
2. Test the resulting program for correctness based on the formal definition.

# MILESTONE REPORT

The milestone report should concentrate on the development of the program. It must include a copy of your formal design documents. The metacognitive portion should address your understanding of the semantics of the constructs.

# FUNDAMENTAL CONCEPTS

Chapter 10 describes the fundamental operations for defining and accessing complex storage structures. The concepts are

1. The amount of storage required must be computed based on the sizing rules of the computer memory system in use.
2. Every allocated memory area has a base address.
3. Every reference is calculated from the base address with an offset.
4. The size of the referenced data is computed by the sizing rules.

In Chapter 10, we develop formulas to calculate the location of an element within an array. The ideas are encapsulated in the C program (shown in Figure VIII.3) with output shown in Figure VIII.1.

The same schema works for structures, except that the computation of the offset is more complicated. As an example, look at the C program shown in Figure VIII.2. From this, we can infer that the offsets computed and then added to the base pointer (char*)(&blspace) is the same as using the more familiar "dot" notation.

```
Element vec[0] has int offset 0
Element vec[0] has byte offset 0
Element vec[1] has int offset 1
Element vec[1] has byte offset 4
Element vec[2] has int offset 2
Element vec[2] has byte offset 8
Element vec[3] has int offset 3
Element vec[3] has byte offset 12
Element vec[4] has int offset 4
Element vec[4] has byte offset 16
```

**Figure VIII.1   Output of Program Processing a Five Element Integer Array.**

```
sizeof(char) 1
sizeof(int) 4
sizeof(int*) 4
sizeof(double) 8
sizeof(struct a1) 16
sizeof(union b1) 16
0 0 0 30ffffff
offset to d: 0
offset to i: 8
offset to c1: 12
```

**Figure VIII.2   Output of struct Text Program.**

## WRITING ARRAYS AND STRUCTURES IN SOL

The SOL syntax makes it easy to develop support for complex data. We can use the type checker to ensure we have the right number of indices as well as taking an interesting view of complex data. To do that, we must make arrays fully functional. First, recall the CRUD acronym for developing data.

### Array Development

1. *Creation.* The creation of an array requires (a) computing the size of the array as discussed in Chapter 10 and (b) allocating the space through the operating system. A straightforward function for this would be [createarray *type dimension*$_1$ ... *dimension*$_n$].

2. *Reading a value from an array.* We have illustrated the concepts of reading a value in both Chapter 10 and earlier in this milestone.

```c
#include <stdio.h>
#include <assert.h>

struct a1 {
  double d;
  int i;
  char c1;
};

int vec[5];

int main(int argc, char* argv[] ) {
  int i;
  int p;
  int *pi, *p0;
  char *ci, *c0;
  for( i = 0; i < 5; i++ ) vec[i] = i;
  p0 = &vec[0];
  c0 = (char*)(&vec[0]);
  for( i = 0; i < 5; i++ ) {
    pi = &vec[i];
    ci = (char*)(&vec[i]);
    /*p = &((char*)vec[i])-&((char*)vec[0]);*/
    p = pi - p0;
    printf(" Element vec[%d] has int
offset %d\n",i,p);
    assert( vec[i] == *(p0+(pi-p0)));
    printf(" Element vec[%d] has byte
offset %d\n",i,(ci-c0));
    assert( vec[i] == *(int*)(c0+(ci-c0)));
  }
}

struct a1 {
 double d;
 int i;
 char c1;
};
```

---

**Figure VIII.3   Class Syntax.**

```
union b1 {
 struct a1 ta1;
 int blast[sizeof(struct a1)/4];
};

int main(int argc, char* argv[] ) {
 union b1 b1_space;
 int i;
 int p;
 printf( "sizeof(char)\t%d\n", sizeof(char));
 printf( "sizeof(int)\t%d\n", sizeof(int));
 printf( "sizeof(int*)\t%d\n", sizeof(int*));
 printf( "sizeof(double)\t%d\n", sizeof(double));
 printf( "sizeof(struct a1)\t%d\n",sizeof(struct a1));
 printf( "sizeof(union b1)\t%d\n", sizeof(union b1));

 for( i=0; i<4; i++) b1_space.blast[i]=-1;
 b1_space.ta1.d = 0.0;
 b1_space.ta1.i = 0;
 b1_space.ta1.c1 = '0';
 printf("%x %x %x %x\n", b1_space.blast[0],
 b1_space.blast[1], b1_space.blast[2],
 b1_space.blast[3]);
 p = (char*)(&b1_space.ta1.d)-(char*)(&b1_space.ta1);
 printf("offset to d: %d\n", p);
 p = (char*)(&b1_space.ta1.i)-(char*)(&b1_space.ta1);
 printf("offset to i: %d\n", p);
 p = (char*)(&b1_space.ta1.c1)-(char*)(&b1_space.ta1);
 printf("offset to c1: %d\n", p);
 assert( b1_space.ta1.d==(*((char*)&b1_space+pd)));
 assert( b1_space.ta1.i==(*((char*)&b1_space+pi)));
 assert( b1_space.ta1.c1==(*((char*)&b1_space+pc1)));
}
```

[structure → *structure-name*
            *optional type declaration*
            *optional* let *statements for variables*
            ... ]

___

**Figure VIII.3   Class Syntax.** (*Cont.*)

Reading a value means (a) computing the offset, (b) computing the address from the base address, and (c) moving the data from the array. The following syntax suffices: [getarray *array dimension*$_1$ ... *dimension*$_n$].

3. *Writing a value into an array.* Writing ("putting") a value into a location of an array is almost identical to reading, except that the new value must be part of the function [putarray *array value dimension*$_1$ ... *dimension*$_n$].

4. *Destroying an array.* Since we use the operating system to allocate the space for the array, it is only natural to use the operating system to deallocate ("destroy") the data—something like [destroyarray *array*].

However, far more interesting is the question of *when* it is safe to destroy such a structure. Java has one answer: garbage collection. We do not pursue garbage collection here, but the serious reader should consider the subject (Jones and Lins 1996). For our application, we leave this decision to the programmer.

### Structure Development

The same line of reasoning and similar functions result from designing support functions for structures: createstructure, readstructure, putstructure, and destroystructure.

Forth has built in structure processing, so the translation of structure functions is less difficult than one might initially estimate. Once again, the issue is computing the offsets.

## CASE 33. Gforth STRUCTURES

Read the ANS specification for Forth. Using this as a guide, develop the semantics of SOL's support for structures.

### Defining Structures

Unlike arrays, structures, by definition, do not have a simple accessing method. Structures are ordered field definitions, *ordered* in the sense of *first, second, ...* . Referencing a data item by its ordinal position would be very error-prone in large structures. Therefore, the usual definition of structures is

1. Start a structure with a keyword. The keyword begins a definitional scope.
2. List the component parts of the structure.
3. Terminate the structure to end the definitional scope.

As an example, C structures look like

```
struct list { double real; struct list *next; }
```

The term `struct` is the keyword and the scope is indicated by the braces. Because structures can be self-referential, the name (`list`) is required in this case.

Creating structures instances is done with `createstructure`, which has the obvious form

> [createstructure *structure-name*]

**Notice** that arrays and structures are types. The introduction of structures requires changes to the type checker to allow for new types. Languages in the ML family also allow for type variables, which greatly enhances the power of the type system.

## Reading and Writing Values

Structures also introduce name ambiguity issues, just as introducing local variables introduced ambiguity. In the case of local variables, the *last defined* rule is used. Such a simplistic rule will not work for structures. Looking at C, we know that if the variable `node` is defined to have type `struct list`, then the expression `node.next` references the second value.

In SOL, the "dot" would get lost next to the left bracket. A different tack is needed. It should be noticed that a structure reference to a field is very much like a function call and the type checker can be used to verify the relation of the field to a data value. Just as with arrays, we introduce two functions: `getstructure` and `putstructure`.

```
[getstructure {\em structure-instance}
                    {\em field name}]\\
[putstructure {\em structure-instance}
                    {\em field name} {\em value}]
```

## Destroying Allocated Structures

Structures that were `allocated` may be `freed` to reclaim space. There is no implied requirement for garbage collection in the SOL specification, although this is an important decision: to have or not to have garbage collection.

# Milestone IX

## CLASSES AND OBJECTS

**Reference Material**

Terrence W. Pratt and Marvin V. Zelkowitz. *Programming Languages: Design and Implementation.* 4th Ed. Upper Saddle River, NJ: Prentice Hall. 1995. Chapter 9.

Kenneth C. Louden. *Programming Languages: Principles and Practice.* 2d Ed. Belmont, CA: Thompson, Brooks/Cole. 2003. Chapters 7 and 8.

### CLASSES AND CLASS

### Objectives

> **Objective 1:** To produce code generation for user-defined classes
> **Objective 2:** To extend the symbol table to include function header information
> **Objective 3:** To give you experience using semantic formalisms used to design semantics
> **Objective 4:** To gain experience in modifying and enhancing your programs
> **Objective 5:** To develop personal project management skills

### Professional Methods and Values

The principles invoked in this milestone are those of the class-oriented paradigm: the encapsulation of names and inheritance.

### ASSIGNMENT

We now have a complete compiler with functions and complex data. All of the development so far makes it possible to implement classes.

## MILESTONE REPORT

The milestone report should concentrate on the development of the program. It must include a copy of your formal design documents.

## ENCAPSULATION AND INFORMATION HIDING

In software engineering, encapsulation means to enclose programming elements inside a larger, more abstract entity. Encapsulation is often taken as a synonym for *information hiding* and *separation of concerns.*

Information hiding refers to a design discipline that hides those design decisions that are most likely to change, protecting them should the program be changed. This protection of a design decision involves providing a stable interface that shields the remainder of the program from the implementation details. The most familiar form of encapsulation is seen in C++ and Java; however, polymorphism is a form of information hiding, also. Separation of concerns is the process of partitioning a program into distinct features. Typically, *concerns* are synonymous with *features* or *behaviors.*

Until the advent of structures, and especially structured programming, in programming languages, all names visible in a scope were visible everywhere in that scope. Structures altered that by making field names invisible within a scope and requiring a special dereferencing operator (the "dot") required. The next step was taken when classes *encapsulated* names, meaning that some names could only be referenced from *within* the class. In this sense, encapsulation is a form of information hiding because `private` names (they could be functions or variables) are a form of design decision that could change if the algorithms change. One step toward objects is easy—make the symbol table associated with a structure unusable *outside* the structure.

## INHERITANCE

In the programming language context, inheritance is a mechanism for forming new classes of data using previously defined classes and primitive data types. The "previously defined" classes are called *base classes* and the newly formed classes are called *derived classes.* Derived classes inherit fields (*attributes*) from the base classes. The attributes include data definitions, sometimes called state variables, and functional definitions, often referred to as methods. An unfortunate asymmetry exists in many object-oriented languages in that primitive data types such as C `int`s do not participate unchanged in inheritance. In these cases, the primitive type is "wrapped" with a class definition.

Inheritance is a method of providing context to state variables and methods. This context is always a rooted, directed acyclic graph (DAG). If only single inheritance is allowed, then the rooted DAG becomes a rooted tree. Examples are obviously Java (single) and C++ (multiple).

## POLYMORPHISM

A consequence of the inheritance hierarchy is that the same method name can occur in many different classes. If all these definitions were in the same visible definition frame, then there would be chaos; one would never know which definition would be used.

This sort of problem actually does exist: overloaded arithmetic operators. Overloading allows multiple functions taking different types to be defined with the same name; the compiler or interpreter "automagically" calls the right one. The "magic" part means that the compiler has a rule for choosing; unfortunately, overloading falls apart quickly. *Parametric polymorphism*, on the other hand, works by matching the function based directly on the types of the inputs. How this works requires an excursion into types.

The type checker, Milestone IV, is not very interesting as type systems go. In adding arrays and structures in Milestone VIII, we still had not deviated much from the C type system. We gain a hint of what is desired by looking at C++'s templates. Actually, the concept of parametric types is quite old, going back to the 1930s. The problem we want to solve notationally is illustrated by the concept of lists. In direct manipulation of lists, the actual internal structure of the elements is irrelevant because we do not—and do not want— to manipulate the information of the elements. Therefore, the actual type of the element is irrelevant and can be anything. This suggests that the type of the elements can be a variable.

This stronger (than C) type theory goes by several names: intuitionistic type theory, constructive type theory, Martin-Löf type theory, or just Type Theory (with capital letters). Type Theory is at once (1) a functional programming language, (2) a logic, and (3) a set theory.

The notation used in the literature is based on the $\lambda$-calculus, except the arguments *precede* the type name. In the theoretical literature, type variables are written as lower-case Greek letters. Using ML, `array` is a type; since we don't have Greek letters on English keyboards, arguments are names preceded by the reverse apostrophe. Here are some ML examples:

| | |
|---|---|
| `'a array` | a one-dimensional array |
| `('a,'b) array` | an array with a pair of types a, b as elements |
| `(int array) array` | an array with integer arrays as elements |

In ML, arrays do not carry their dimension as part of the type, which is determined at runtime. ML does not have a separate notation for array indexing, either. Despite this, ML arrays are useful data structures just as C.

The useful nature of type variables allows us to express data structure types algebraically, leading to work in category theory and algebraic specifications.

# PART II

## GENERAL
## INFORMATION

The purpose of this part is to develop an understanding of the issues relating to designing and implementing a programming language. Chapter 4 is a project to implement a very simple language.

# 4

## PROJECT

## 4.1 CLASS

### 4.1.1 Week 1

The first class day is spent with the usual administrative tasks. The deliverable for the first day is to get personal information from the students and perform what I call "expectation management." Unless your students have significant experience with problem-based learning and other active learning pedagogies, it is important to explain the process. One exercise that you might consider is some form of "ice-breaker" exercise; there are many suggestions on the World Wide Web.

Time and time management are an important aspect of the course. Our counseling center has prepared one-hour lectures on studying issues, including time management. The chapter on PDSP has forms and information, but how to actually develop time plans is best explained by instructors with experience.

The first exercise for the course is the introduction to the specification, Chapter 2. I read the students into the project by telling them that the role I play is that of a liaison with the language development group; I am not here to make decisions or dictate solutions. My job is to guide. There are two deliverables for the first week: (1) a small suite of test programs and (2) the beginnings of the lexicon of compiling.

For active-learning techniques to work, the students must be actively engaged. My experience is that this is doubly threatening to students: (1) they lose their anonymity and (2) they come under active scrutiny during discussion. This second point is the more difficult because in any discussion there are questions and answers that are tossed around. This is called the Socratic method or the Moore method. My experience is that students feel if you ask a follow-up question, then their original answer was wrong. You will have to constantly remind them that the Socratic method is an approach that requires each participant to ask questions and to also answer them.

It is helpful to get the students thinking about the technical vocabulary and the relationships among words. I have found mind-mapping exercises to be helpful.

### 4.1.2 Week 2

The second week is taken up with understanding how humans process and understand language. Even though SOL is a "formal language," it is a language first and we must develop a vocabulary for describing features of language. This vocabulary is traditionally part of the philosophy of language: linguistics. The key issue is that the students must begin to separate syntax, semantics, and pragmatics and develop formal understanding of the deeper issues.

The deliverable for week two should be the global design of the system. Among the resource texts, one can find a single diagram that shows the components of a compiler and the gross processing. An example is in Pratt and Zelkowitz (1995), Figure 3.2, page 80. At the end of the week, each component should have gross processing requirements developed and data structures named.

## 4.2 MIND MAPS

### 4.2.1 The Exercise

Mind maps are the result of free word association. Mind mapping is a graphical technique for exploring the relationships among words. The procedure is simple.

Take a piece of paper (11 × 17 computer paper is excellent). Turn the paper long-side down (landscape mode). Write the word that you want to map in the center. Now, as rapidly as you can, write all the words you can think of that the subject word reminds you of. It doesn't matter what the relationship is—all you care is that the subject word reminds you of the new word. Now, the new word may remind you of another word... go ahead, follow that path. At some point, the original subject word will come back into view, so to speak. Start a new leg using that word.

Continue this process as long as it seems fruitful. After doing this exercise a few times, you will learn when you're done. As untechnical as it sounds, you will run out of new ideas—that point will be obvious to you.

Mind mapping is a "brainstorming" technique.

### 4.2.2 Why It Works

That this exercise is fruitful is taken as evidence for how human memory functions. Psychologists believe that concepts are related in memory in a network fashion. This means there are many different connections between concepts. When we try to remember something, we search our memory along all the connections that exist among related words. The human brain is quite slow when compared to the modern computer. Nonetheless, this search method can go on without conscious control; this is why you may suddenly remember something long after you gave up consciously searching for the connections.

# 5

---

# INTRODUCTION TO LINGUISTICS

Programming languages are, first and foremost, languages. Languages have been studied by philosophers for several thousand years. It seems appropriate to begin our study of programming languages by reviewing the more fundamental ideas of language.

Language is fundamentally associated with representation, communication, meaning, and truth. These same issues arise in programming languages, although they are rarely discussed in the literature.

Modern studies of language attempt to study language in a systematic way, often using mathematical formalisms. A group known as *logical positivists* that included Bertrand Russell, Ludwig von Wittgenstein, and Rudolf Carnap was particularly influential. The positivists are important because of their profound impact on computer science. The positivists strove to employ rigorous accounts of logic and of meaning in attempts to penetrate, and in some cases to dispel, traditional philosophical questions. The positivists sought complete meaning by investigating language through *linguistic analysis.*

## 5.1 FOCUS PROBLEM

The purpose of this chapter is to develop the technical vocabulary of linguistics and language. The "deliverable" for this chapter is an initial decomposition of the term "compiler."

### 5.1.1 Technical Vocabulary

The philosophy of language (linguistics) introduces a set of standard, technical concepts that can be applied to any language, natural or formal. All languages have *syntax, semantics,* and *pragmatics* (see Figure 5.1).

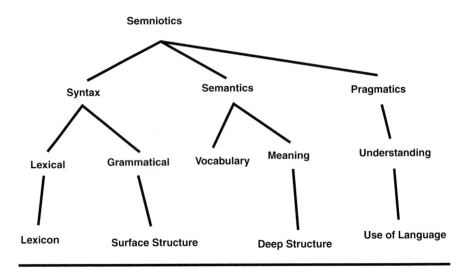

**Figure 5.1   Linguistics Word Association Map**

What are these properties? How do they describe elements of a given language? What subproperties do they each imply? In order to develop a compiler, we must first have an understanding of the problem *from the problem-poser's standpoint.*

### 5.1.2   Deliverable

By the end of the chapter, we need to have a written understanding of what the basic parts of the compiler should be, what its data should be, and what the overall processing sequence should be.

The method for this deliverable is the following:

1. Read the chapter. Make a list of all words that relate to language and linguistics.
2. Classify each word as to its part of speech: noun, verb, etc.
3. Choose a simple expression from our list of examples.
4. Write a simple story of how the example is processed, using the following rules:
   - Common nouns are the names of data types.
   - Proper nouns are variable names.
   - Verbs are the names of processes.

Suppose we choose [let [a 1]] as our example. The story should explain how the line is taken from the file, parsed, and coded. Obviously, at this point, there will be many details lacking, but some very basic information gained.

## 5.2  PHILOSOPHY OF LINGUISTICS

Natural (human) languages are powerful and complex systems. The science of this capacity and of these systems is *linguistics*. The term *computational linguistics* generally means studying natural language computationally. We adopt, where necessary, the more descriptive term *programming language linguistics* as the study of languages used to describe computations, which may include descriptions meant for other than the computer. Like other sciences, and perhaps to an unusual degree, linguistics confronts difficult foundational, methodological, and conceptual issues.

We seek systematic explanations of a language's syntax (the organization of the language's properly constructed expressions, such as phrases and sentences), its semantics (the ways expressions exhibit and contribute to meaning), and its pragmatics (the practices of communication in which the expressions find use). This study is called *semniotics*.

The study of syntax has been guided since the 1960s by the work of Noam Chomsky. Chomsky, who lends his name to the Chomsky Hierarchy in theoretical computer science, takes a cognitivist approach. Human linguistic capacities, he holds, are innate and come from a dedicated cognitive faculty whose structure is the proper topic of linguistics. Chomsky's direct contribution to programming languages—indeed, a fundamental contribution—comes from a 1957 article concerning a hierarchy of syntactic structures. In programming languages, syntax is specified using the principles of *formal languages* in general and context-free languages in particular, largely in line with Chomsky's ideas. The fundamental contribution of formal grammars and production systems is the work of Emil Post.

Semantics, on the other hand, is far more difficult. Ideas of programming languages are received primarily from *logic* and *recursive functions*. Here many of the great strides have been made by philosophers, including Gottlob Frege, Alonzo Church, Bertrand Russell, Ludwig von Wittgenstein, Rudolf Carnap, Richard Montague, Saul Kripke, Dana Scott, and Christopher Strachey to name but a few. The role of semantics in programming language linguistics is development and careful application of formal, mathematical models for characterizing linguistic form and meaning.

Philosophical interest in pragmatics typically has had its source in a prior interest in semantics in a desire to understand how meaning and truth are situated in the concrete practices of linguistic communication. As an example of the role of pragmatics, consider the computational concept of *increment an integer variable* i and the C statement i = i+1. More than one person has commented that "this is a ridiculous statement since no integer is equal to its successor." We can immediately see that programming language pragmatics is its own study.

Modern programming systems use many different input media, not just keyboards. Therefore, we should look broadly at *language* and investigate semniotics, which is the study of signs and signification in general, whether linguistic or not. In the view of the scholars in this field, the study of linguistic meaning should be situated in a more general project that encompasses gestural communication, artistic expression, etc. For example, mathematical notation constitutes a language in just this way.

## 5.3  MEANING: LANGUAGE, MIND, AND WORLD

Language, of course, is not a subject in isolation (see Crimmins 1998). On one hand, language is used for communications, and on the other it is used to represent thoughts in the mind. The relationships among means of communication constitute the meaning of language.

Language may be studied from many directions, but we are interested in how language relates to two "communities": one is the programmer community and the other is the computer development community. Yes, this is a bit of whimsy, but in its essence, programming languages are about transferring information from the human programmer to the computer. In this act, the compiler of the programming language must be aware of the communication aspect as well as the preservation of intention.

In some real sense, meaningfulness in programming language derives from the mental content and intent of the programmer, perhaps including the contents of beliefs, thoughts, and concepts. This means there is a cognitive aspect to understanding programming languages.

It has not gone unquestioned that the mind indeed can assign meaning to language; in fact, skepticism about this has figured quite prominently in philosophical discussions of language. It has also played a large role in developing programming languages. There are three basic approaches to assigning meaning to programming language constructs.

1.  Operational. Operational meaning was the first meaning theory based on the naïve idea that the computer evaluation of a construct was the meaning of the construct. But what if one changes computer systems?
2.  Axiomatic. Axiomatic theories of meaning are based on specifying the connotation of a construct by logical means. *Connotation* means that an essential property or group of properties of a thing is named by a term in logic.
3.  Denotational. Denotational semantics was proposed as a mechanism for defining meaning by specifying a mapping from syntactic constructs to values. Often called Strachey–Scott semantics for its proposers, denotational semantics is a part of $\lambda$-calculus and recursive functions. Denotation is a direct specific meaning (value) as distinct from an implied or associated idea in connotation.

It is important to understand the role of language in shaping our thoughts. The Sapir–Whorf hypothesis states that there is a systematic relationship between the grammatical categories of the language a person speaks and how that person both understands the world and behaves in it. While the stronger versions of this hypothesis are no longer believed by researchers, it is clear that having language is so crucial to our ability to frame the sophisticated thoughts that appear essential to language use and understanding, that many doubt whether mind is *prior* to language in any interesting sense.

One must be careful about the relationship of language and the world. According to this picture, the key to meaning is the notion of a truth-condition. The meaning of a statement determines a condition that must be met if it is to be true. According to the truth-conditional picture of meaning, the core of what a statement means is its truth-condition—which helps determine the way reality is said to be in it—and the core of what a word means is the contribution it makes to this.

While the truth-conditional picture of meaning has dominated semantics, some philosophers urge that the key to meaning is a notion of correct use. According to this alternative picture, the meaning of a sentence is the rule for its appropriate utterance. Of course, the two pictures converge if sentences are correctly used exactly when they are true. The challenge illustrates a sense in which the Mind/Language and Language/World connections can seem to place a tension on the notion of meaning (meaning is whatever we cognitively grasp, but the meaning of language just is its bearing on the world).

Our first task in understanding programming languages is to understand the terminology used to describe such languages. That terminology is itself a language. These early cases will guide you to an understanding of how humans process language; that process is the basis for developing a program to process and translate language.

## CASE 34. WHAT DOES AN INFINITE LOOP MEAN?

For most computer science students, all this background is new information and you might ask, "Why should I care?" In order to convince you, the reader, that all is not as simple as we would like, write a short paper on the following question:

What is the meaning of the C statement `while(1);`?

## 5.4   WHAT IS SYNTAX?

When first approaching a problem that we have never solved before, we are in a learning mode. We must effectively learn enough about the underlying concepts of the problem so we can form a coherent picture of the problem.

A fundamental tenet of learning theory is that we must first determine what we actually know. We start out by investigating two things we do know about: we all speak natural languages and we all can program.

For this portion of the course, we need to understand language in the broader context of natural language. This provides us with important anchor points. As we discover what we know about natural language, we can ask, "Well, how does this particular idea play out in programming languages?"

Since you have made it this far in your education, you have undoubtedly studied at least one natural language. You may be lucky enough to speak several languages. While you can learn a language without being completely aware of technicalities, we must now understand those technicalities because discussing programming languages requires these technical issues.

We are interested in invariant parts of language and we call these structures. The term *structure* in this context means "the mutual relation of the constituent parts or elements of a language as they determine its character; organization, framework."[*]

*Syntax* is a term that denotes the study of the ways in which words combine into such units as *phrase, clause,* and *sentence.* The sequences that result from the combinations are referred to as *syntactic structures.* The ways in which components of words are combined into words are studied in *morphology*; syntax and morphology together are generally regarded as the major constituents of grammar (informally, grammar is synonymous with syntax and excludes morphology). Syntactic descriptions do not usually go beyond the level of the sentence.[†]

*Morphology* is the study of the structure of words, as opposed to syntax, which is the study of the arrangement of words in the higher units of phrases, clauses, and sentences. The two major branches are inflectional morphology (the study of inflections) and lexical morphology (the study of *word formation*).[‡]

---

[*] *The Oxford Dictionary of English Grammar,* Ed. Sylvia Chalker and Edmund Weiner. Oxford University Press, 1998. Oxford Reference Online. Oxford University Press. Clemson University. 12 February 2005 <http://www.oxfordreference.com/views/ENTRY.html?subview=Main&entry=t28.e1424>

[†] *Concise Oxford Companion to the English Language,* Ed. Tom McArthur. Oxford University Press, 1998. Oxford Reference Online. Oxford University Press. Clemson University. 12 February 2005 <http://www.oxfordreference.com/views/ENTRY.html?subview=Main&entry=t29.e1200>

[‡] *Concise Oxford Companion to the English Language,* Ed. Tom McArthur. Oxford University Press, 1998. Oxford Reference Online. Oxford University Press. Clemson University. 12 February 2005 <http://www.oxfordreference.com/views/ENTRY.html?subview=Main&entry=t29.e804>

*Grammar* is "the system by which words are used together to form meaningful utterances. It denotes both the system as it is found to exist in the use of a language (also called descriptive grammar) and the set of rules that form the basis of the standard language, i.e., the variety of a language that is regarded as most socially acceptable at a given time (also called prescriptive grammar)." *

*Lexical* structures are those relating to lexicography, "the principles and practices of dictionary making. Belonging to, or involving units that belong to, the lexicon. E.g. a lexical entry is an entry in the lexicon; a lexical item or lexical unit is a word, etc. which has such an entry; rules are lexically governed if they apply only to structures including certain lexical units."†

The purpose of this chapter is to explore these definitions and to understand how they play out in programming language.

## CASE 35. WHAT CAN WE LEARN FROM A DICTIONARY?

The central concept in *vocabulary* is the dictionary. The dictionary in a compiler is generally called a *symbol table,* although there may, in fact, be many different tables, depending on the designer's view. Central to understanding the symbol table is understanding how an English language dictionary is laid out. Study an unabridged dictionary, such as *Merriam-Webster* or the *Oxford English Dictionary.* Develop a data representation for the *structure* of the dictionary and outlines for the following *queries:*

1. Looking up a word to determine whether or not it's in the dictionary
2. Looking up a word when you want to know the definition within a specific part of speech
3. How to handle multiple definitions of the same part of speech

This case provides a good opportunity to describe *mind maps* as a brainstorming device. Consider Figure 5.2. The concept behind mind mapping is word or phase associations. In the figure, the word "Dictionary" was placed in the center of the palette.‡ The right-hand side of the map was drawn by considering the words or terms that are suggested by the

---

\* *Pocket Fowler's Modern English Usage,* Ed. Robert Allen. Oxford University Press, 1999. Oxford Reference Online. Oxford University Press. Clemson University. 12 February 2005 <http://www.oxfordreference.com/views/ENTRY.html?subview= Main&entry=t30 .e1674>

† *The Concise Oxford Dictionary of Linguistics.* P. H. Matthews. Oxford University Press, 1997. Oxford Reference Online. Oxford University Press. Clemson University. 12 February 2005 <http://www.oxfordreference.com/views/ENTRY.html?subview= Main&entry=t36 .e1719>

‡ The FreeMind software, available at SourceForge, was used for this example.

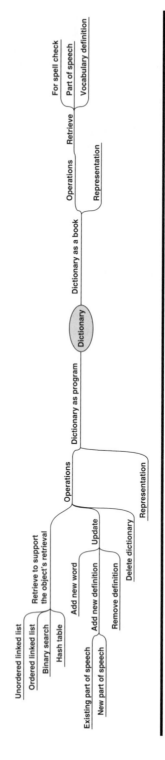

**Figure 5.2 Mind Map of the Dictionary Case**

word dictionary. In this case, the searching functions came to mind. The left-hand side was derived in the same way, but with the information of the right-hand side clearly in mind. The left-hand side also makes use of the CRUD mnemonic for developing data structures: **C**reate, **R**ead, **U**pdate, and **D**estroy. These are the four issues that must be addressed in design.

## CASE 36. HOW DOES WHAT WE LEARNED PLAY OUT IN PROGRAMMING LANGUAGES?

The preceding case should have given you some ideas about describing the symbols of a language. A dictionary entry is a word often with a large number of meanings. Each word also has a part of speech. Common words often have many meanings that are quite different in different contexts.

We might expect to find context in our study. *Context* is "the speech, writing, or print that normally precedes and follows a word or other element of language. The meaning of words may be affected by their context. If a phrase is quoted out of context, its effect may be different from what was originally intended."*

Forgetting the contextual issue for the time being, our next task is simply to identify all the parts of speech that commonly occur in current programming languages. Let's be specific.

*http://java.sun.com/docs/books/jls/second_edition/html/j.title.doc.html*

is the Java Language Specification. (A more reliable way to find this is to search for "Java Language Specification" in a search engine.) Your assignment is to develop a list of categories of symbols that appear in Java. If you spend some time in the Table of Contents, you will find exactly what you need to know in one place. For some ideas on how to read technical literature, see Section 5.7 on Focused SQRRR.

You must develop a table with the following columns:

| Category | Use | Spelling Rules |
| --- | --- | --- |

It turns out that the spelling rules are well modeled by finite state machine diagrams. The specification provides enough detailed information to make this easy.

---

* *Concise Oxford Companion to the English Language*, Ed. Tom McArthur. Oxford University Press, 1998. Oxford Reference Online. Oxford University Press. Clemson University. 12 February 2005 <http://www.oxfordreference.com/views/ENTRY.html?subview=Main&entry=t29.e309>

## 5.5 GRAMMAR

## CASE 37. ENGLISH GRAMMAR WITH PRODUCTION RULES

The purpose of the preceding cases is to develop a vocabulary about words. Natural language sentences consist of words, but not just any arbitrary order. Natural language grammars specify the order of words, but it is not that simple.

English is organized in a hierarchy of types: sentences, clauses, phrases, and words. Sentences are made up of one or more clauses, clauses consist of one or more phrases, and phrases consist of one or more words. From our preceding work, we know that words are classified by their part of speech, so we might substitute *part of speech* for *word*. Also, the hierarchy is not just a hierarchy because a clause can appear in a phrase, for example.

Our purpose is to understand how grammars are developed and used so we can develop a similar idea in programming language. The mechanism often used is to develop a syntax tree that represents the relationship of the hierarchy in the sentence at hand.

I will describe the *thinking process*. Let's consider a *simple sentence*.

*My brother is a lawyer.*

This is a simple sentence that I denote as *S*. A simple sentence is defined to consist of a *noun phrase (NP)* followed by a *verb phrase (VP)*. Symbols such as *NP* are called *categorical symbols* because they stand for whole sets of possible sequences of symbols. One way I can describe this rule is to invent some simple notation to write with. I choose to use a notation similar to that used in formal languages in computer science.* I will encode the sentence: "A simple sentence is a noun phrase followed by a verb phrase." by

$$S \rightarrow NP\ VP$$

The *application* of that rule now divides the sentence *S* into two pieces. To indicate this relationship, I will rewrite the sentence, collecting the various pieces. This rewriting process is called a *derivation*: the *rule* (column 2) is applied to the input (column 1) and the result then becomes input to the next step:

---

* This isn't surprising because computer science originally took the notation from linguistics in the 1950s and 1960s.

| Step | Input String | Rule Applied |
|------|--------------|--------------|
| 1 | [S: My brother is a lawyer] | S → NP VP |
| 2 | [S: [NP: My brother] [VP: is a lawyer]] | |

The verb phrase "is a lawyer" is the next thing to look at. A verb phrase is a verb ($V$) followed by a noun phrase

$$VP \rightarrow V NP$$

| Step | Input String | Rule Applied |
|------|--------------|--------------|
| 1 | [S: My brother is a lawyer] | S→ NP VP |
| 2 | [S: [NP: My brother] [VP: is a lawyer]] | VP → V NP |
| 3 | [S: [NP: My brother] [VP: [V: is] [NP: a lawyer]]] | NP → ADJ N |
| 4 | [S: [NP: [ADJ My] [N: brother]][VP: [V: is] [NP: [ADJ: a][N: lawyer]]]] | |

where we use the rule (twice) that a noun phrase can be an adjective followed by a noun.

Each of the bracketed units is a word, a phrase, or a clause. We refer to these as *constituents*, defined as a word or a group of words that act syntactically as a unit.

Although the brackets are able to portray the grammatical hierarchy, this notation is cumbersome. If we are to make use of the ideas of grammars, then we need to have something more computable. It turns out that *hierarchy* is a synonym for *tree* and trees are something computer scientists know how to deal with. A tree for the statement "My brother is a lawyer" based on our grammar is shown in Figure 5.3. The tree is drawn with two types of arcs between nodes. The solid arcs are *grammatical*, or what is known as the *deep structure*. The dashed arcs indicate *bindings* of specific words to grammatical *terminals*. A *binding* is an association of a term to its definition. The power of the grammar is that a countable number of sentences can be developed by binding different words to the terminals; for example, "Our house is a home." Meaning is derived from the deep structure.

## CASE 38. GRAMMARS IN PROGRAMMING LANGUAGES

We have completed exercises that explore morphology and we know something about grammar. We need to finish syntax considerations by looking at the Java grammar. You should have seen that the Web site document contains the actual grammar that can be used to define Java. This grammar is larger than the simple example above, but it is exactly the same idea.

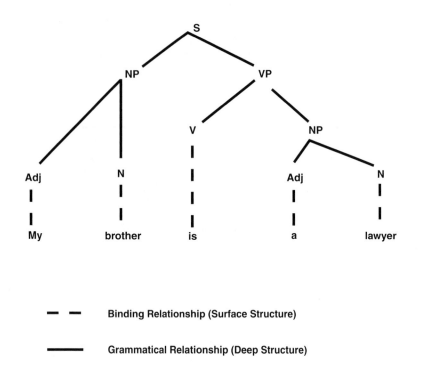

**My brother is a lawyer**

— — Binding Relationship (Surface Structure)

——— Grammatical Relationship (Deep Structure)

**Figure 5.3   Syntax Tree for "My Brother Is a Lawyer"**

If you have forgotten the details of what a grammar is or how it is specified, then you can read Chapters 1 and 2 of the Java Specification.

Here is a simple Java program:

```
public static void main(String args[]) {
        System.out.println(''Hello World'');
    }
```

Chapter 16 has the full Java grammar. Your task is to start with "MethodDecl" and follow it until you have derived a simple function.

The basic points for Case 38 can be portrayed by a concept map. A concept map is similar to a mind map but there is one very important difference. The nodes in a concept map are concepts, but two concepts must be connected by a relationship. In the concept map shown in Figure 5.4, the concepts are enclosed in figures; the relationships have no enclosing

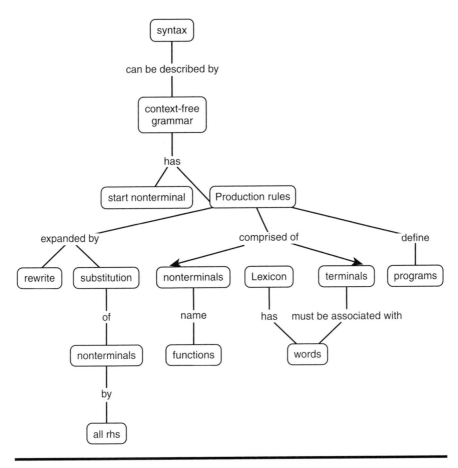

**Figure 5.4   Concept Map for Syntax and Grammars**

lines. Concept maps are directed, bipartite graphs. This particular concept map was drawn by software available from the Institute for Human and Machine Cognition at the University of West Florida.

## 5.6   SEMANTICS

To this point we have only been concerned with the *form* of the sentence. We have not made much of a distinction between the written and the spoken sentence. Because programming languages are often written (but need not be: consider visual programming languages) and the project certainly is, the discussion is slanted toward written information. Based on the discussion in Case 37, it is clear that sentences in either mode can be written in a tree wherein the leaves are the exact words and the interior nodes represent production names. While this tree represents a good bit of the

| Word | Part of Speech | Usage |
|------|----------------|-------|
| My | adjective | modifies *brother* |
| brother | noun | subject |
| is | verb | verb |
| a | article | modifies *lawyer* |
| lawyer | lawyer | predicate noun |

**Figure 5.5   Usage of Words in "My Brother Is a Lawyer"**

information needed to understand a sentence, it is not the whole picture. How can we investigate this process and turn it to our understanding of programming languages?

*Semantics* is the study of meaning and this is much different from syntax. Because graphical methods are desirable in the computing context, it would be nice to have a mechanism to consider semantics from a graphical point of view. One mechanism is *sentence diagrams*. The purpose of a sentence diagram is to demonstrate the relationships among the words of a sentence, as well as the purpose of groups of words. For example, the case of: "My brother is a lawyer" has the usage pattern shown in Figure 5.5. A diagram of this sentence, along with its parse tree, is shown in Figure 5.6.

1. *brother* is the *subject*.
2. *is* is the *verb*.
3. *lawyer* is the *predicate*.
4. *my* is an *adjective* modifying *brother*.
5. *a* is an *adjective* modifying *lawyer*.

## 5.7   FOCUSED SQRRR

Focused SQRRR stands for "Focused SURVEY, QUESTION, READ, RECITE, REVIEW (FSQRRR)." FSQRRR is a technique for reading when there is a specific question to be answered. Again, the purpose is to minimize reading material that does not bear on the problem at hand.

In FSQRRR, we start with *focus questions*. The focus question may be given to you by your instructor or it may be a question that you have formulated in trying to solve a problem. The question must be written out before you start!

### 5.7.1   Survey and Question

When using this form, write the questions in the spaces provided. Remember that this is a working document: brevity is fine as long as clarity does not suffer.

My brother is a lawyer

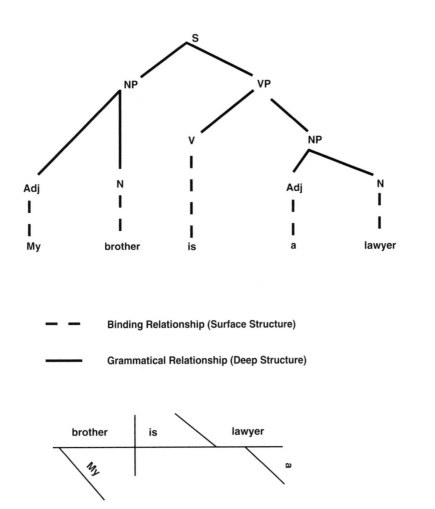

**Figure 5.6 Syntax Tree and Diagram of Sentence**

1.  Write the focus question. Now, rewrite it using your own words. If the question contains technical terms you do not understand, then put the term in parentheses but then write out what you think the term means.
2.  Read the title and turn it into a question that is suggested by the focus question.
3.  Read the introduction and formulate questions related to the focus question. If the section introduction indicates that this

section does not bear on the focus question, come back to it only if you later believe you need to. (For example, the section that actually addresses the focus question might rely on information developed in this section.)

4. Turn headings and subheadings into questions as explained above.
5. Read captions under pictures, charts, and graphs again, looking for relevance to the focus question.
6. Read the summary. The summary should give you an outline of how the chapter helps answer the focus question.
7. Recall the instructor's comments.
8. Recall what you already know. This is especially crucial with respect to vocabulary.

### 5.7.2 Read

Before reading further, rewrite the focus question in light of what you have learned in the SQ portion.

1. Look for answers to your questions.
2. Look for answers to the teacher's questions and the book's questions.
3. Reread the captions.
4. Study graphic aids.
5. Carefully read italicized and bolded words.
6. Reread parts that are not clear.

### 5.7.3 Recite

Now try to put in your own words what you think the answer to the focus question is.

### 5.7.4 Review

Within 24 hours of your research:

1. Page through material and reacquaint yourself with important points.
2. Read written notes you have made.
3. Reread the focus question and rewrite it in light of what you now know.
4. Reread your answer to the focus question. Rewrite it.

# 6

---

# LINGUISTICS FOR PROGRAMMING LANGUAGES

**Reference Materials**

Terrence W. Pratt and Marvin V. Zelkowitz. *Programming Languages: Design and Implementation.* 4th Ed. Upper Saddle River, NJ: Prentice Hall. 1995. Chapters 1 and 3.

Kenneth C. Louden. *Programming Languages: Principles and Practice.* 2d Ed. Belmont, CA: Thompson, Brooks/Cole. 2003. Chapter 1.

*Web Sites*

Wikipedia: search "metalanguage," "surface structure," "deep structure," "transformational grammar."

docs.sun.com

## 6.1   METALANGUAGE VERSUS OBJECT LANGUAGE

Speaking about languages adds a strange twist to the conversation. How do you distinguish between the language you are talking about and the language you are using to describe features? For example, when you learned a programming language such as C, you probably used a natural language such as English to describe a feature of the language C. In this case, we use the natural language as a metalanguage while C is the object language. The distinction is important because during the project, there are several languages in use: the production language for the parser, the type definition language for types, Forth, and Gforth. We must be very clear as to which (if any) is the object language and which is the metalanguage.

A metalanguage describes an object language. The prefix "meta-" in the case at hand denotes a language that deals with issues such as behavior, methods, procedures, or assumptions about the object language. The object language is the language system being described.

This distinction becomes evident when reading programming language specification. This chapter uses the Java specification as an exemplar for language specifications. The specification is available at `java.sun.com`.*

## 6.2 SURFACE STRUCTURE AND DEEP STRUCTURE

When you read natural language, there are certain features of the language, such as word order or spelling, that we recognize as belonging to that language. These features we call *surface structure*. For example, native English speakers would probably question the word "al Asqa" as being an English word. But there is another structure: a *deep structure*. Deep structure deals with the manner in which lingustic elements are put together in a sentence. In Chapter 5 we developed the syntax tree of a sentence and we discussed the idea of sentence diagrams; both of these concepts are deep structures.

## CASE 39. WHAT IS THE PROPER WAY TO FORM JAVA SYMBOLS?

In Chapter 5, Section 5.4, we introduced the concept of morphology: the structure of words. A synonym for morphology is **lexical structure**, which is also related to the concept of surface structure. Certainly, programming languages have many sorts of words, including numbers and quasi-mathematical symbology. How should we define the spelling rules for a language. We can study a well-developed language such as Java to find out.

For this case, refer to Chapter 3, "Lexical Structure of the Java Language Specification," in (name a book?). Please note that you may first have to read Chapter 2, "Grammars," to familiarize yourself with the manner in which production rules are portrayed. Answer the following questions:

1. What is **unicode** and why is it important? Is unicode just arbitrary numbers? Could a Russian programmer use a cyrillic keyboard for Java?
2. Read the chapter and develop a lexicon of lexical concepts in programming languages. For example, **floating point numbers** and **identifiers** are two such types. In this table, you should have major headings (such as "Numbers") and subheadings (such as "Integer").

---

* Accessed in April 2005.

## CASE 40. WHAT ARE THE PARTS OF SPEECH IN JAVA?

Read Chapter 18, "The Grammar of the Java Programming Language," in (name of book?) What is the equivalent to the parts of speech in natural language? **Hint:** Link the lexical categories to parts of speech as terminals for the grammar.

## CASE 41. HOW IS THE SYNTAX OF JAVA PRESENTED TO DEVELOPERS

It is time to consider how we communicate the technical aspects of the syntax of a programming language. Again consider Chapter 2 and Chapter 18 of the Java specification.

Starting with the nonterminal *statement* develop a parse for the statement

```
a = b+c;
```

## CASE 42. GRAPH MODELS OF SYNTAX

One of the major reasons for using the context-free type of grammar is that we can develop a very usable data structure: *an n*-ary tree or a directed acyclic graph (DAG). The procedure for developing such a graph is the following.

Suppose we have the following context-free production

$$N \to X_1 \ X_2 \ \dots \ X_n$$

where $N$ is a nonterminal and the $X_i$ are either terminal or nonterminal. We can form a *rooted, ordered n-ary tree* by making the root of the tree, and the results of the parse of $X_1$ is the first child, $X_2$ is the second child, and so on.

Using the parse of the previous case, draw the tree.

## 6.3   STRUCTURAL INDUCTION

The fundamental approach to designing semantics is called *structural induction*. Consider once again Old Faithful:

$$E \to T + E \mid T - E \mid T$$
$$T \to F * T \mid F/T \mid F\%T \mid F$$
$$F \to I \in \text{int} \mid (E)$$

How many different *forms* can be generated from this grammar? The short answer is "There are an infinite number of possible forms." The naïve answer would say that $1 + 2 * 3$ and $7 + 5 * 12$ are different forms: surface forms!

How many forms from the deep structure? There are just nine: one for each right-hand side of the productions. Thus, we can build up, or more properly *induce*, the structure from the productions. For this reason, the technical name for this method is called *structural induction* and a companion approach for recursion, structural recursion.

## CASE 43. STRUCTURAL INDUCTION

This method is central to our understanding of many applied and theoretical algorithms. Write a research paper on structural induction and structural recursion. For cases, a research paper is intended to be one or two pages long. The focus questions in this exact instance are, "What is a definition of each term?" and "Demonstrate that you understand by producing one or more examples." In this paper, give a concrete problem (such as string generation) and illustrate the use of structural induction to generate possible answers. Similarly, develop a recursive definition of the same problem and demonstrate structural recursion.

## 6.4   INTERPRETATION SEMANTICS

The purpose of a program is to produce *values*. In order to have the computer do useful work we must, in a stepwise fashion, instruct the computer on exactly how values are to be computed. You are somewhat familiar with this idea from computation in mathematics. In the early grades, we work with numbers and operations only. We are the computer in this circumstance and we are the ones who must learn to add, subtract, multiply, and divide. As we become more advanced, we learn to compute values of trigonometric functions and logarithms. When we learn calculus, we learn how to compute with symbols using the rules of differential and integral calculus.

In each case, we learn three things: (1) how to formulate problems, so that we can (2) write down computations so that they are understood by everyone, and (3) how to convert the symbols into values. We use mathematical notation in, for example, calculus class, that has evolved over a 400-year time span. Programming languages are much younger (1950s), but they fulfill the same purpose as mathematical notation: a system of symbols (language) that express algorithms (programming). In the case of programming languages we can write algorithms that compute symbols as well as numbers. One such class of programs is *compilers and interpreters* that read programs and cause the algorithms to be executed. In order to do this, we must agree on various words and constructs are converted to computer instructions. We call this agreement *semantics*.

We have discussed semantics in Chapter 5, where we defined it as the study of meaning of words, phrases, sentences, and texts. The same general definition applies here, except we are thinking about a *formal, mathematical* meaning. Here's an example: the mathematical statement

$$3 + sin(2\pi)$$

can be carried out by the Forth statement

3 2 PI * fsin + if PI has been set equal to a proper approximation to $\pi$:

3.14159265358979323846 according to the IEEE standard.

What then is the *semantics* of 3 2 PI * fsin +? We will take the view in this text that the semantics of programming language is the value computed by the expression, or the *denotational semantics*. Denotational semantics are based on the $\lambda$-calculus. The $\lambda$-calculus is discussed in Chapter 9.

There are several approaches to language, denotational being one. *Axiomatic semantics* is developed in the same manner as any axiomatic (mathematical) system. Axiomatic semantics are closely associated with Edgers Dijkstra, David Gries, and Sir C. A. R. Hoare. This approach is perhaps informally known to you because it introduces the concepts of *precondition* and *postcondition*. The fundamental inference rules are formulated by

$$\{P\} S\{Q\},$$

where $P$ is the precondition, $Q$ is the postcondition, and $S$ is the command. In axiomatic semantics, the semantics name not the value (as in operational and denotational), but the conditions under which the values exist, the expression's connotation. An informal use of this concept is seen in testing because test cases require the tester to understand the three parts of the formula.

An *operational semantics* is a mathematical model of programming language execution that utilizes a mathematically defined *interpreter*. This mathematically defined interpreter is actually in the spirit of the definition of Gforth. The first operational semantics was based on the $\lambda$-calculus. Abstract machines in the tradition of the SECD machine are closely related. SECD was introduced by Peter Landin in 1963. His definition was very abstract and left many implementation decisions to the implementors. This situation gave rise to research by Plotkin (1981) and later investigators.

## 6.5   WHAT ABOUT PRAGMATICS?

Finally, then, how do programming languages display pragmatics? Actually, the answer is simple: the pragmatics are seen in the structure of programs. Recall that "pragmatics" means the manner in which the language is used. Simply, then, programming language pragmatics are the elements of programs and the learned idioms. For example, the English language statement "increment i by one" is pragmatically displayed in C in several ways: i++, i+=1, and i=i+1. These are semantically equivalent but may still be pragmatically different because they are used in different circumstances.

# 7

## WHAT IS A GENERAL STRUCTURE FOR COMPILING?

---

**Reference Materials**

Terrence W. Pratt and Marvin V. Zelkowitz. *Programming Languages: Design and Implementation.* 4th Ed. Upper Saddle River, NJ: Prentice Hall. 1995. Chapters 1 and 3.
Kenneth C. Louden. *Programming Languages: Principles and Practice.* 2d Ed. Belmont, CA: Thompson, Brooks/Cole. 2003. Chapters 1 and 3.

---

## 7.1 A GENERAL FRAMEWORK

We have been studying language, both natural and formal. Formal languages are languages that have completely specified syntax and semantics. What we believe at this point is that programming languages are not all that different from natural languages in their overall concepts. We should expect that programming languages have the same parts (Figure 7.2): although syntax deals with the overall appearance of the language, the meaning of the words is the purview of semantics; informally semantics has taken on the meaning of "meaning of a phrase or sentence." Regardless, there must be a lexicon and a vocabulary. The lexicon is the entire set of words, whereas the vocabulary is words and definitions. We also have studied semantics in terms of its relationship to the deep structure of a statement.

From a project management standpoint there are some questions: who, what, when, where, why, and how are each of these different parts defined?

## CASE 44. MERGING LINGUISTICS AND COMPILING

Write a story about how the sentence, "The cat ate the hat." is understood using the phases listed in Figure 7.1. In linguistics and logic, we speak of the binding time of a word. The binding time is the moment when the definition of a word is assigned. In an abuse of language, we can extend this concept to relate to the assignment of meaning for any element: word, phrase, sentence, paragraph, .... When are the meanings of the five words bound? When is the meaning of the sentence bound?

Meaning is not something that just happens: it occurs as we read the sentence. We can only understand the meaning of the sentence when we understand the meaning of the underlying words, phrases, etc. The same must happen in the compiler.

## CASE 45. PROCESS ORDERING

Develop a time line representation of the translation of a C statement x = 1*2+3 into the Forth statement 1 2 3 * +. Annotate the time line with the phases in Figure 7.1. When is the lexicon consulted? How is the vocabulary used?

## 7.2   TYPES

When you study the assembly language for a computer (not just the chip) it is apparent that the vast majority of instructions manipulate various data types: integers, floating point numbers, information to and from devices, among others. It would be difficult to move a program from one computer to another if there were not some organization to this data and some

 

(1)  Syntax

    (a)  Morphology

        (i)  Spelling rules
        (ii)  Lexicon
        (iii)  Vocabulary
    (b)  Grammar and deep structure
(2)  Semantics
(3)  Pragmatics

**Figure 7.1   Phases in Compiling**

agreement on what various operations are called. This organization is called *data types* or just *types*.

In order to explain types, let's look at the quintessential type: integers. Everyone knows about integers; everyone knows that "+" is read "plus" and stands for the operation "take two integers and sum them giving an integer as a result." Everyone also knows that "<" is read "is less than" and stands for the condition "one number is less than another number, giving a true or false answer." Everyone also knows that "if $a$ is less than $b$ and $b$ is less than $c$, then $a$ is less than $c$."

For our purposes, we can start by defining a *type* as a triple of the form $\langle S, F, R \rangle$, where $S$ is a set of constants, $F$ is a set of function symbols, and $R$ is a set of relation symbols.

$$\text{Integers} = \langle integers, \{+, -, \ldots\}, \{=, \neq, <, \ldots\}\rangle$$

It should be clear that $S$, $F$, and $R$ determine the lexicon for integer expressions (without variables). The harder part is to develop algorithms for each element in $F$ and $R$.

## CASE 46. PRIMITIVE DATA TYPES

Develop a type definition for each of the standard types: booleans, integer, floating point, and strings.

## 7.3   RECAP

In this chapter we have laid the groundwork for the project. We have a time line of who, what, when, where, and why elements of a statement are processed and given meaning. This is the fundamental step in any project. We are missing the *how* for many of these steps. The purpose of the Milestone chapters in Part I is to fill in the *how* and to implement a simple compiler that takes simple statements to Forth.

## 7.4   DEFINING A LANGUAGE

In order to understand the requirements in the milestones, we need to understand the overall issues in a programming language.

1. Architectural
   - Registers
   - Control word
   - Control flow
   - Input/output
2. Primitive data
   - Arithmetic—integer and floating point
   - Logical (bit level)
   - Character
   - Array support
3. Pseudo-instructions
   - Value naming
   - Relative addressing
   - Static storage allocation

**Figure 7.2    Classes of Statements in Assembler Language**

### 7.4.1    Algorithms and Data Representation

Fundamentally, a programming language is a vehicle for describing algorithms and data representations. In the early days (1950s and 1960s), the major implementation vehicle was assembly language. Assemblers make algorithms and data representations quite clear because only primitive machine data types are available. The major categories of statements are shown in Figure 7.2.

You can see that we still have those same classes of statements. While hardware is faster and with more memory, the same classes of statements are present in today's chips.

### 7.4.2    Evolution

Why did programming languages evolve? The simple answer is that programming large systems in assembler language is very costly and very error prone. One of the first steps was to eliminate the deep connections between the program and the architecture. Programming at the assembler level is by definition tied to a specific computer on which the implementation is done. This causes a major problem: the code must be changed if there is any change in the underlying machine. The ramifications were far reaching in the 1960s: manufacturer dominance was equivalent to software system dominance. This stifled innovation in the hardware arena because the cost of recoding a system rarely could be justified by the advantages of new hardware except for pure speed.

By 1960 there were several higher-level languages that abstracted the computer from the algorithm; the most influential were ALGOL, Cobol, Fortran, Lisp, and Snobol. Fortran remains heavily used in scientific programming and Lisp continues to be important in artificial intelligence applications. Object-oriented concepts emerged in 1967 with Simula67. Therefore, by 1970, the stage was set for evolution in many directions.

Higher-level languages effectively removed the architectural class by having variables and data controlled by the compiler. Some languages continue to have input/output defined in the language, but most have now eliminated that from the language definitions and moved to using a library of subprograms instead.

Looking at the pragmatics of programming language development, there has been a steady movement from complex languages (PL/I) and smaller libraries to smaller languages with much larger library support (JAVA and its API). The one major exception has been Fortran, which has always had extensive library support due to its use in numerical processing. For example, C was develop almost directly in revolt against the complexity of PL/I and Pascal. The C/Unix pair, developed at Bell Laboratories in the late 1960s and early 1970s, became the standard in computer science education; C's minimalistic nature influenced many language designers.

Newer languages such as ML that have significant complexity introduce expanded function declaration mechanisms, such as *modules*. Modules encapsulate operations for entire data types, for example. Objects, on the other hand, encapsulate data types along with memory allocation.

### 7.4.3   Implications for the Project

Regardless of the exact design of a language, we will be able to develop and categorize various saspects of the language in a manner similar to that shown in Figure 7.3. Each and every aspect will have an impact on all the possible phases of the compiler.

## CASE 47. DESIGNING MODULES

Use the form given in Figure 7.4 to develop the details needed. Some language elements may have entries in all columns, some may not. **Note:** The semantics column must reference a Forth word in the table and that word must be defined below the table.

1. Architectural
   - Control flow
     a. if-then-else
     b. for / do
     c. while
     d. subprograms: functions and subroutines
   - Input/output subprograms
2. Primitive data
   - Arithmetic—integer and floating point
   - Logical (bit level)
   - Character and strings
   - Arrays
   - Pointers and dynamic allocation
   - Structures and objects
3. Pseudo instructions
   - Variables and assignment statements

**Figure 7.3   General Classes of Statements in Higher-Level Languages**

|         | Syntax  |            | Type | Semantics | Pragmatics |
|---------|---------|------------|------|-----------|------------|
| Lexical | Grammar | Vocabulary |      |           |            |

**Figure 7.4   Design Table for Language Features**

Each symbol with *semantic* meaning must be in the table. From a practical standpoint, it is probably best to start with the various primitive data types and operations.

# 8

---

# A UNIFIED APPROACH
# TO DESIGN

---

**Reference Materials**

ANSI/IEEE X3.215–199x. Draft Proposed American National Standard for Information—Programming Languages—Forth. X3J14 dpANS–6—June 30, 1993.

Terrence W. Pratt and Marvin V. Zelkowitz. *Programming Languages: Design and Implementation.* 4th Ed. Upper Saddle River, NJ: Prentice Hall. 1995.

Kenneth C. Louden. *Programming Languages: Principles and Practice.* 2d Ed. Belmont, CA: Thompson, Brooks/Cole. 2003.

Lawrence C. Paulson. *ML for the Working Programmer.* Cambridge University Press. 1996.

*Online Resources*

www.gnu.org: GNU Foundation. *GForth Reference Manual*

www.Forth.org

---

## 8.1  INTRODUCTION TO DESIGN

Designing a large system is much different from designing a single object or function. The close relationship of object-oriented design methods and object-oriented programming languages, for example, is no mistake: programming languages, software engineering, and design have co-evolved throughout the history of computer science.

However, there are many other disciplines that use design principles and therefore other approaches. This chapter outlines the use of what has been learned in psychology about problem solving. Although there seems to be a dearth of texts in computer science about design, such books abound in the engineering sciences, architecture, and the trades. One approach adopted in engineering is to treat design as problem-solving. This approach is not unknown in mathematics (Pólya 1957) and computer science (Simon 1969).

## 8.2   TWO ELEMENTS

Design of a large project requires a great deal of organization. Anecdotal evidence from industry indicates that a large project such as a compiler requires three to five years. Another piece of anecdotal evidence is that an experienced programmer produces 2.5 debugged lines of code per hour. With such a long development time, documentation and coordination are paramount concerns. Many documentation methods and approaches have been proposed but there is no silver bullet (Brooks 1975).

The approach here is based on concepts of problem solving as proposed by psychologists. The ideas are relatively simple and easy to document but they're still no silver bullet. The two elements are called *concept maps* and *schemata*.

### 8.2.1   Linguistic Issues

A *concept* is a word that names a class of objects, real or imagined. Concepts play a central role in epistemology, the theory of knowledge. Concepts are often thought to be the meanings of words or, perhaps better, the names of the meanings of words. Thus, a concept is a single word that can stand in for many different categories. A *category* is a set of objects. For example, the concept written "city" names a number of possible categories: "city" or "metropolis" or "town," but not "village." This idea of concept and of category is very important to us during design. The idea of *class* in object-oriented design is the same as our idea of category.

Concepts are essential to ordinary and scientific explanation: someone could not even recall something unless the concepts they hold now overlap the concepts they held previously. Concepts are also essential to categorizing the world; for example, recognizing a cow and classifying it as a mammal. Concepts are also compositional: concepts can be combined to form a virtual infinitude of complex categories, in such a way that someone can understand a novel combination. For example, "programming language" can be understood in terms of its constituents as a "language for programming machines." However, if one understands "language" but not "programming," then "programming language" is meaningless.

The set of all words that we can use in discussing a project is its *lexicon*. The lexicon can be used to develop the *vocabulary*. The vocabulary is the set of pairs of a lexicon word and its *definition*. In order to understand design we must understand the concepts and relationships among them. The concepts and relationships come from the specification and this is almost always presented in a graphical form. We therefore must first undertake a linguistic analysis of the specification. Linguistic analysis provides us with the lexicon of the project and the vocabulary.

The above examples show that concepts can be related with one another. A well-known example in artificial intelligence and object-oriented programming is the "is-a" relationship. The "is-a" relationship is a variant of Aristotle's genus-species rules: "A tree is a graph."

So there you have it: concepts, relationships, lexicon, vocabulary. Concepts are related to other concepts based on stated relationships *in the context of the discussion at hand*. These relationships can be graphed as a DAG. Why DAGs? Because we cannot afford to have cycles in definitions!

## 8.2.2 Schemata

Concept maps portray factual knowledge, but such knowledge may not be static. The concept "bachelor" as an unmarried adult male is a fact of the definition and is unlikely to change. On the other hand, the concept of "democracy" is not so trivial. Concepts play an important role in the design because they set the range of assumed knowledge: the facts of the specification.

Design in computer science has a unique requirement. The product must always be a program written in a programming language. In order to do that, you must do the following:

1. Recognize the concepts in the requirements.
2. Determine the semantics of the problem in the *problem* vocabulary. This is the concept map.
3. Determine the semantics that must work in the programming language.
4. Determine the syntactic constructs to encode the semantics.

This is part of what makes programming hard: the translation of semantics to semantics, not syntax to syntax.

We present here one possible way to understand this process using a concept central to the psychological process of problem solving: the *schema* (sing.) or *schemata* (pl.). A schema has four interconnected parts:

1. A set of patterns that can be matched to circumstances
2. A set of logical constraints and performance criteria
3. A set of rules for planning
4. A set of rules for implementing

This idea actually originated in early artificial intelligence research in the 1960s. Rule-based programming systems such as CLIPS were developed based on the schemata concept. Concepts play a role in the first two items. We spend the rest of the chapter giving examples of these ideas.

A very simple example would be the following problem:

Write a program in C to compute the sum of the first $N$ cubes.

The linguistic analysis is easy: *C program* is obvious; *cubes* refer to *cubes of integers*. The phrase *compute the sum of* immediately indicates a loop of some sort. In this case, the "loop of some sort" is a schema.

1. The pattern that we recognized was the need for the loop based on the statement of the problem.
2. There is an obvious logical requirement that deals with the counter of the loop. There is a subtle question here; do you see it?
3. There is some planning that must be done for a loop: in this case, $N$ must be known to control the loop. Where does it come from? And what do we do with the output? The input requirement (for $N$) invokes several schemata: we could read it in from a file or read it on the command line or some other element. The output also invokes several possible schemata.
4. When all the planning is done, we can write the loop.

Did you consider the question in point 2? How big can $N$ be? The summation of cubes runs as $N^4$. You can look that up in a mathematics handbook; the key point was to realize that there is a logical requirement on $N$. For a 32-bit signed-magnitude architecture, only $N < \sqrt[4]{2^{32}} = 2^8$ computes properly.

## 8.3  USING THESE IDEAS IN DESIGN

What does this have to do with compiler design? The design of the transformations is effectively schemata.

1. The input is a series of elements that have linguistic meaning.
2. The input must meet a required pattern. The input can be broken into a series of patterns through decomposition. The name of each schema comes from the grammar.
3. Each pattern, when recognized, causes a standard processing sequence to be invoked. The processing is recursive, but the number and content of each pattern schema are known. After imposing logical requirements and planning requirements (which is usually a decomposition of sorts), the transformation can be processed.

$$E \rightarrow T + E \mid T - E \mid T$$
$$T \rightarrow F * T \mid F/T \mid F\%T \mid F$$
$$F \rightarrow I \in \text{int} \mid (E)$$

**Figure 8.1    The Old Faithful Grammar**

How does this work in practice? Here's the famous Old Faithful grammar that recognizes integer arithmetic expressions (Figures 8.1 and 8.2). Figure 8.1 is really shorthand for the grammar shown in Figure 8.2. Recall from the discussion of grammars in Chapter 5 that the elements on the left-hand side are called nonterminals and any symbol that does not appear on the left-hand side is called a terminal. Why these terms are used will become apparent below.

By convention, $E$ is called the sentential symbol: every possible string of symbols (called a sentence) made up of integers and the operations symbols must be constructible beginning with just the $E$. Although the form of the rules may appear strange, they can be converted to a program (see below).

### 8.3.1    Understanding the Semantics of Grammars

In order to make sense of this schematic form of programming, let's invent a simple interpreter that takes programs written as above and executes them. This exercise is crucial: it shows exactly what we must design to produce a compiler.

$$E \rightarrow T + E$$
$$E \rightarrow T - E$$
$$E \rightarrow T$$
$$T \rightarrow F * T$$
$$T \rightarrow F/T$$
$$T \rightarrow F\%T$$
$$T \rightarrow F$$
$$F \rightarrow \text{any integer}$$
$$F \rightarrow (E)$$

**Figure 8.2    Unrolled Old Faithful Grammar**

All the rules for a given nonterminal can be gathered into one C function by developing rules to follow when translating one to the other. In our case, we only need to go from grammars to programs; that you can go the other way is covered in the proof of the Chomsky Hierarchy theorem.

### 8.3.1.1  Reading Terminals

It is clear that the terminal symbols appear verbatim on the input. We can assume that there is a function called `read_symbol()` that reads the next well-formed symbol from the input if there is one; it returns EOF if there is no more input. It will be useful to have a function `next_symbol(string)` that returns *true* if the next input symbol is that string and *false* otherwise.

### 8.3.1.2  Dealing with Nonterminals

There are two ways we can deal with the nonterminals:

1. We can look on the input and determine what nonterminals the input might have come from. This process can be made very efficient, but it is not particularly intuitive. Such an approach is called *bottom up*.
2. We can start at the sentential symbol and *predict* what should be there. This is not as efficient because in general one has to search a tree (bottom up does not in general do this) although there are techniques to prevent this. This search technique is called *top down* or *recursive descent* when no backup can occur.

We will take the recursive descent approach because it leads to a more intuitive design process. Remember, you can't optimize a nonfunctioning program. We're working for understanding here, not speed.

Working for $F$ upward, how would we program a C program to do $F$'s functioning?

$$F \rightarrow \text{any integer}$$
$$F \rightarrow (E)$$

In keeping with the theme of schemata, what is the most general processing schema we can think about? The illustration here uses the *input–process–output* schema. Every algorithm you will ever write will have those elements. That's interesting, but what is the schema for a program? It has

- A name
- Some arguments represented by names and defined by types
- Some processing constructs using `if`, `while`, and so forth
- Some method of returning values or perhaps writing results

```
returntype programname(arguments) {
        Processing
        Return
}.
```

**Figure 8.3   A Program Schema for Programs**

We can see that the first two follow from the *input* requirement and the last one from the *output* requirement. The C schema for this is something like that shown in Figure 8.3.

How would we map the productions for *F* onto this schema? Clearly, *F* is the name, so let's call it progF. What is the input? That is not clear so we will wait on this. What does it return; it may not be clear but it needs to return a tree (Figure 8.4).

In terms of our schema model, we have pattern matched things we know from the problem to the schema. Let's focus on the *Processing* part. It is clear that there can only be two correct answers to *F*'s processing: either we find an integer or we find a left parenthesis. Anything else is an error—or is it? What about an expression such as (1 + 2)? This causes us to consider a different issue: that of *look-ahead*. Put that issue on a list of issues to be solved.

There is another issue for *F*. If *F* has a left parenthesis, how does it process the nonterminal *E*? The rule we want to have is that each non-terminal is a function and therefore recognizing a string requires *mutual recursion* among several functions. The thought pattern is "If *F* sees a left parenthesis, then call progE with no arguments. When progE returns, there must be a right parenthesis on the input."

## 8.3.2   Designing a Postfix Printer Program

To illustrate how to use the grammar to design, let's develop a program that prints an expression in Polish postfix. There are two issues: (1) how

```
tree progF( ??? ) {
     tree Result
     Processing
     return Result
}
```

**Figure 8.4   Schema for progF**

```
match expression {
      case pattern: code
           ⋮
      default: code
}
```

**Figure 8.5  Match Schema**

to translate the grammar form above into a C program and (2) what is the starting point?

Let's invent a new program element, called match, that syntactically looks like a C switch statement. The general form is shown in Figure 8.5.

Furthermore, let's assume that someone has actually implemented the match command in C. This is not that far fetched because functional languages of the ML-based variety all have such a facility. In order to organize this we turn each *nonterminal* into a function name, for example, *E* into progE (Figure 8.6). Each function has one argument, Tree.

Everything in this program except the match-case are regular C. Try your hand at completing the functions printT and printF before considering the hand calculator program.

To review, the use of the match-case construct encodes the schema. The case construct is a pattern. Because the program is so simple, we did not have a separate criterion, planning, or implementation phase. However, the printT and printE can be thought of as planning steps. The putchar function is clearly an implementation step. Also, there is no knowledge to apply to the schema.

```
void printE( tree Tree ) {
    match Tree
    case T+E: printT(T); printE(E); putchar('+'); return;
    case T-E: printT(T); printE(E); putchar('-'); return;
    case T: printT(T); return;
    default: abort();
    }
}
```

**Figure 8.6  Program to Print *E* Recognized Terms**

```
int progE( tree Tree ) {
    match Tree
    case T+E: return (progT(T)+progE(E))
    case T-E: return (progT(T)-progE(E))
    case T : return progT(T)
    default : abort()
    }
}

int progT( tree Tree ) {
    match Tree
    case F*T: return (progF(F)*progT(T))
    case F/T: return (progF(F)/progT(T))
    case F%T: return (progF(F)%progT(T))
    case F : return progF(F)
    default : abort()
    }
}
```

**Figure 8.7  Complete Design for Print Program**

## CASE 48. MATCH-CASE SEMANTICS

Write a one- or two-page paper that outlines *exactly* how the match-case statement should operate.

### 8.3.3  Calculator

One of the classic examples of this form of programming is the program that acts like a hand calculator. This example appears in the so-called "Dragon Book" of Aho, Sethi, and Ullman (1986); it is called the "Dragon Book" because of the distinctive cover featuring a dragon.

In the hand calculator program, we take in any expression that is syntactically correct by our grammar and this expression is a tree. The purpose is to compute the integer result. You should know from your data structures experiences that the expressions matched by Old Faithful grammar can be represented in a tree (see Figures 8.7 and 8.8). **Note**: What are the logical requirements for correct computation based on the limits of representation? Can you check these *a priori?* How would you find out the limits have been violated?

```
int progF( tree Tree ) {
    match Tree
    case I: return atoi(I)
    case(E): return E
    default : abort()
    }
}
```

---

**Figure 8.8   Complete Design for Print Program (continued)**

## 8.4   DESIGN RECAP

The design of a compiler follows from the material in Chapters 5 to 8. The principles are

1. The input text is converted to *tokens* that capture the type of grammar terminal.

```
void progE( tree Tree ) {
    match Tree
    case T+E: progT(T); progE(E); print("+")
    case T-E: progT(T); progE(E); print("-")
    case T : progT(T)
    default : abort()
    }
}

void progT( tree Tree ) {
    match Tree
    case F*T: progF(F); progT(T); print("*")
    case F/T: progF(F); progT(T); print("/")
    case F%T: progF(F); progT(T); print("%")
    case F : progF(F)
    default : abort()
    }
}
```

---

**Figure 8.9   Complete Design for Gforth Program**

```
void progF ( tree Tree ) {
      match Tree
      case I: printf(I)
      case(E): progE(E)
      default : abort()
      }
}
```

**Figure 8.10 Complete Design for Gforth Program (continued)**

2. The tokens are converted by the parser into a tree that captures the deep structure of the program.
3. The trees can be processed, sometimes multiple times, to produce structures that describe the sequence of operations.
4. The final structure can be converted to the program run by a computer, or, in our case, Gforth.

As a final example, let's convert the calculator program into Gforth code (Figures 8.9 and 8.10).

# 9

---

# TECHNICAL FOUNDATIONS IN PROGRAMMING LANGUAGES

---

### Reference Materials

Terrence W. Pratt and Marvin V. Zelkowitz. *Programming Languages: Design and Implementation.* 4th Ed. Upper Saddle River, NJ: Prentice Hall. 1995. Chapter 4.

C. Hankin. *Lambda Calculi: A Guide for Computer Scientists.* London: King's College Publications. 2004.

Kenneth C. Louden. *Programming Languages: Principles and Practice.* 2d Ed. Belmont, CA: Thompson, Brooks/Cole. 2003. Chapters 9 and 13.

### Web Sites

Wikipedia: search "lambda calculus"

---

## 9.1 INTRODUCTION

Early models of programming languages were very ad hoc simply because there was so little accumulated experience and research. In the early 1960s, the predominant theoretical model of computation was the Turing machine, the model received from logic and mathematics. The Turing machine made sense as a model because computers of the day "looked" similar. Interestingly enough, there were many other programming models available.

## CASE 49. EARLY HISTORY OF PROGRAMMING LANGUAGES

There are many interesting Web sites on the history of programming languages, but the more interesting study is that of computing from the time of Charles Babbage's original analytical engine (1819) through 1970 when there was an explosion of programming paradigms.

Write a paper that includes a timeline that marks events in the development of programming languages/machines/systems. For each event:

1. What was the major innovation? What was the problem that demanded this innovation?
2. Where did the innovation come from and from whom?
3. How does the innovation enhance computing?

Merge your timeline with others in the class to form the class's perception of the history of innovations.

## 9.2 BINDING TIMES: WHEN DO WE KNOW?

We probably don't think about it, but history plays an overarching role in the evolution of language. We pick up the current Java release manual and don't think much about how we got it. That's unfortunate, because we lose perception of the workings.

The term *binding time* is used to denote when, along the development timeline, a word in the lexicon is bound to (associated with) its definition. As an example, consider the program snippet $\sin(x)$ and ask "When do $\sin$ and $x$ receive their values?" The use of the abbreviation $\sin$ for the sine of an angle is inherited from mathematical practice many centuries ago. The actual code for computing the sine is relatively recent but is based on the Taylor series, named for Brook Taylor (1685–1731). The code itself (which is a definition) was most likely written in the language itself, but it is a system routine and the regular programmers treat it as available. What about $x$? $x$ is invented by the programmer during program design and long after the compiler is written. Assuming it is not a constant, it received its definition (value) during the running of the program. Therefore, stretched out on a timeline, the binding times for such a simple snippet stretch to antiquity through the program's running.

The reason this discussion takes places is that it is the prelude to the discussion of *scope*.

## 9.3 SCOPE

Scope is the term that denotes the rules under which a symbol can be replaced by a definition. For students with only C-based language experience (C, C++, Java), scope rules are relatively straightforward. This is because of the relative inability to introduce new procedures. The functional languages tend to have much richer definitional capabilities.

The basic issue in scope is the implementation of a *visibility function*. An example of how this issue affects us is the following simple C program:

```
int mystery(float x){
double xp = x;
  {
  int xp = 17;
  return xp;
  }
}
```

What is returned? Why?

The metaphor of *seeing* is very real here: the `return` statement sees the *nearest* definition of xp and hence returns the 17. How can the compiler deal with this?

The issue is *environment-producing statements*, such as the braces in C. An innovation in the 1970s with structured programming was the realization that all programs can be written with properly nested environments. An environment is a set of all the definitions that have occurred in the scope of the environment producing statements.

In mystery, there are *three* environments. The program header/name is in the file environment. The x and the first definition of xp are in the second and the last definition of xp is in the third environment. The return statement is in the third environment, so it uses the definition of xp in that environment. What about this problem?

```
int mystery(float x){
double xp = x;
  {
      statements using and changing xp;
      return xp;
  }
}
```

In this case, there are still three environments, but no definition of xp occurs in the third. Therefore, the `return` statement uses the definition of xp from the second environment. This is why the proper nesting of definitional environment is so crucial.

## CASE 50. SOL DEFINITIONAL ENVIRONMENTS

Using the SOL specification, identify all the environment-producing statements of the language. Develop several examples of how the environments can be nested; these examples should serve as test cases in implementation.

# 9.4 THE λ-CALCULUS MODEL

All programming languages have the ability to define independent procedures. The complication comes when a procedure wants to use another procedure—or even itself—to compute values. As straightforward as it seems today, the concept of how to think about variables, arguments, and invocation was not always spelled out. The first and most influential was the idea of the λ-calculus, first developed by Alonzo Church in the 1920s, and later refined in *The Calculi of Lambda-Conversion* in 1941. Because of Church's interests, the calculus was developed strictly over the integers; naïve extensions to other types is straightforward, although a complete theoretical explanation had to wait until Barendregt in 1984.

More importantly for our purposes, the λ-calculus was the basis of the programming system Lisp, primarily motivated by John McCarthy, which gained prominence in the 1960s. McCarthy wrote "A Basis for a Mathematical Theory of Computation" in 1963, which had the major impact of moving the basis of programming from Turing machines to recursive functions as founded on λ-calculus. The Lisp 1.5 manual in 1965 presented a programming system we would recognize: Lisp is coded in Cambridge Polish. What follows is a simple introduction to the λ-calculus suitable for our purposes.

## 9.4.1 Syntax

A suitable syntax for λ-expressions would be Old Faithful with just one or two additions. For reference,

$$E \rightarrow T + E$$
$$E \rightarrow T - E$$
$$E \rightarrow T$$
$$T \rightarrow F * T$$
$$T \rightarrow F/T$$
$$T \rightarrow F\%T$$
$$T \rightarrow F$$
$$F \rightarrow \text{any integer}$$
$$F \rightarrow (E)$$

### 9.4.1.1 Variables

The first task question is to define the lexical rule for variable. The obvious answer is that we use elements composed of letters. In the original formulation, variables were single, lower-case Latin letters.

Where should variables be recognized? A moment's thought indicates that variables are stand-ins for integers. Therefore, we need a new rule for $F$:

$$F \rightarrow V: \text{any lower-case Latin letter}$$

### 9.4.1.2 Introducing Variable Names

The numeral 17 is converted to the same number everywhere in a program, but, as we discussed above while dealing with scope, a single variable has meaning only in scope. How do we indicate scope? Old Faithful has no means; in fact, neither did arithmetic in general until Church came along.

Church introduced an *improper function* $\lambda$ to introduce a new variable. For example, $\lambda\ x\ .\ x + 2$ introduces the variable $x$. Notice the period: the expression on the right of the period is the expression "guarded" by the $\lambda$.

There is no obvious place in the existing grammar to place the $\lambda$, but it is clear that the right-hand side could be any complex expression. Therefore, we add a whole new nonterminal and we'll call it $L$:

$$L \rightarrow \lambda V: \text{any variable}. E$$

An obvious problem is that $F$ has a rule for $(E)$; this should be changed to $(L)$. In order to have parenthesized expressions on the right-hand side of the period, we need a renaming production for $L$:

$$L \rightarrow E$$

Finally, there is no way to define a function. Adopting a common format we use the let keyword:

$$letprod \rightarrow \text{let} V: \text{any variable} = L$$
$$\rightarrow E$$

## 9.4.2 Semantics

Consider the following two statements in the extended calculus:

$$let f = \lambda\ x.x + 2\, f(3)$$

What is the value of $f(3)$? To answer this we must define the semantics of the $\lambda$-calculus. Once the $\lambda$ is taken care of, the semantics are those of ordinary arithmetic.

The semantics we chose to demonstrate are those of **substitution**, **rewriting**, and **unfolding**.

### 9.4.3 Substitution and Rewriting

The concept of substitution should be clear, at least in the naïve sense. Restrict the substitution operation to one variable only at a time. Substitution would be a three-place function that defines what it means for a value $v$ to be substituted for a variable $x$ in expression $e$. As sample semantics, look at $f(3)$: it should be clear that we want to take the 3 and substitute that value everywhere $x$ occurs on the right-hand side of the $\lambda$-expression. In order to keep track of this substitution, we need to **rewrite** the expression. We propose that the semantics of the $\lambda$ is this substitution/rewrite action. We can display that in a tabular form:

| Old Expression | New Expression | Reason |
|---|---|---|
| $f(3)$ | $(\lambda x.x + 2)(3)$ | Substitute definition of f |
| $(\lambda x.x + 2)(3)$ | $3 + 2$ | Definition of $\lambda$ |
| $3 + 2$ | 5 | Definition of $+$ |

### 9.4.4 Recursion and the Unfolding Model

Consider a slightly different definition: factorials.

$$let\ ! = \lambda x.x * f(x - 1)$$

Actually, this would be a nonterminating computation unless we have some method of checking for a zero argument. It should be obvious by now how to add an "if" statement to Old Faithful so that

$$let\ g = \lambda x.\ if\ (x = 0)1\ else\ x * g(x - 1)\,f(3)$$

This is clearly much different from the first function we defined. It also adds a complication that has plagued theoreticians for 100 years: the $g$ that is defined is not necessarily the $g$ on the right side of the expression. Programmers recognize this same problem when they are required to use definitions in C to cause the loader to correctly match the definition. There have been several different solutions to this problem: a special form of the `let`, usually termed a `letrec`, is common. Another solution is to require that every `let` is actually `letrec`. Let's use this latter idea.

The semantics for recursion are effectively the same, substitution and rewriting, but with some wrinkles. The tabular presentation is wasteful of space, so we will resort to a simplification.

$$g(3) \Rightarrow (\lambda x.\ \textit{if}\ (x=0)\ \textit{then}\ 1\ \textit{else}\ x*g(x-2))(3)$$

$$\Rightarrow \textit{if}\ (3=0)\ \textit{then}\ 1\ \textit{else}\ 3*g(3-1)$$

$$= 3*g(2)$$

$$\Rightarrow 3*(\lambda x.\ \textit{if}\ (x=0)\ \textit{then}\ 1\ \textit{else}\ x*g(x-1))(2)$$

$$\Rightarrow 3*(\textit{if}\ (2=0)\ \textit{then}\ 1\ \textit{else}\ 2*g(2-1))$$

$$= 3*(2*g(1))$$

$$\Rightarrow 3*(2*(\lambda x.\ \textit{if}\ (x=0)\ \textit{then}\ 1\ \textit{else}\ x*g(x-1)))$$

$$\Rightarrow 3*(2*(\textit{if}\ (1=0)\ \textit{then}\ 1\ \textit{else}\ 1*g(1-1)))$$

$$= 3*(2*(1*g(0)))$$

$$\Rightarrow 3*(2*(1*(\lambda x.\ \textit{if}\ (x=0)\ \textit{then}\ 1\ \textit{else}\ x*g(x-1))))$$

$$\Rightarrow 3*(2*(1*(\textit{if}\ (0=0)\ \textit{then}\ 1\ \textit{else}\ g(0-1))))$$

$$= 3*(2*(1*1))$$

$$= 3*(2*1)$$

$$= 3*2$$

$$= 6$$

where $\Rightarrow$ indicates a substitution move and $=$ indicates a rewrite move.

We can now state what compilers actually do: they generate code that, when executed, carries out the subsitution, rewrite, and unfolding operations. The question before the compiler designer is how to do this efficiently.

## 9.4.5 Arguments, Prologues, and Epilogues

All this looks very easy when you can work out a small problem on paper, but in large systems, the bookkeeping is very difficult. Actually, the code for computing values is reasonably straightforward compared to all the bookkeeping: getting values to the called functions, remembering which function was the caller and where to return, how to store values so they can be retrieved easily, and how to reclaim space when it is no longer needed.

### 9.4.5.1 Argument Passing

Consider the following snippet:

```
void caller( ...){

int x;

...

a = called(3 + 3.7 − x);

...

}


float called ( double x ){

int q,z;

...

return (q + 3.0)/z;

}
```

This problem shows many of the issues in actually implementing a programming language with regard to functions, control, and argument passing.

1. The argument $3 + 3.7 − x$ must first be evaluated.
2. The result must be stored appropriately.
3. Control must be transferred to `called` with enough information so that `called` can return the value to the `caller`.
4. `called` must save information regarding the state of `caller`.
5. `called` then computes the requested information.

6. called then must restore caller's state and put the return value where caller can find it.
7. called must return control to caller.
8. caller must store the return value into a.

There are many possible protocols that can be used for these steps. Generally, the tasks cover the two steps starting at 3, called the **prologue**, and the two steps starting with 6, called the **epilogue**.

## CASE 51. ARGUMENT PASSING

The actual protocol for how arguments are passed is many and varied. Write a short paper on how these protocols have been handled in the past. Choose from the list the protocol you want for SOL.

## 9.5   IMPLEMENTATION ISSUES

This chapter has presented a number of concepts that must be defined in order to produce a compiler. Here is a partial list of tasks that must be finished before a design is completed.

1. Give an operational definition of the visibility function.
2. How are you going to allocate and free storage for arguments?
3. How are you going to allocate and free storage for local variables?
4. What are the scope-producing statements in SOL? When does the scope start and when does it end?
5. How will values be returned?
6. How will control be transferred from the calling procedure to the called procedure?

# 10

STORAGE MANAGEMENT
ISSUES

Regardless of type or lifetime, values must be stored in computer memory during their lifetimes. The purpose of this chapter is to address the issues of designing and implementing storage schemes.

## 10.1  COMPUTER MEMORY

Each manufacturer has its own detailed implementation; most present a von Neumann architecture to the programmer (Figure 10.1). Memory tends to be passive; information is stored at and retrieved from locations determined external to the memory. Again using the CRUD acronym to guide development, the active agent must

> Create: a place in memory that is not shared by any other storage unit
> Read: values by copying and not destroying content
> Update: content without interfering with other storage units
> Destroy: the content so that the space can be reused

Most commonly available computers can address memory on the byte (8-bit) level, although word (32-bit) or double precision (64-bit) is not unknown. For our discussion here, let's define *primitive storage unit* as the number of bits the memory transfers (either read or write).

Any element that can be represented in a single primitive storage unit is easy to control; however, almost all data, including primitive data types such as integers, take more than one primitive storage unit. Virtually all operating systems provide primitives that `allocate` and `deallocate` memory. In general, `allocate` has one argument, the number of primitive storage units requested and returns a memory address. To return the memory to the free pool, `deallocate` takes a previous `allocated` memory address

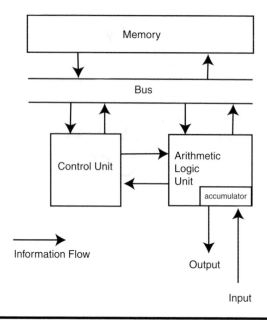

**Figure 10.1    Scheme of von Neumann Architecture**

and that area of memory is released. Therefore, the CRUD for memory is as follows.

- Create is a call to `allocate`.
- Read is the *dereferencing* of a pointer.
- Update is assigning new information on a pointer.
- Destroy is a call to `deallocate`.

Understanding this is the key to understanding space issues in compilers. For every type, the compiler must

1. Determine the number of primitive storage units required to hold the information in the type
2. Provide for the allocation of the space before it is needed
3. Maintain access and update control of the space during its lifetime
4. Deallocate the space at the end of its lifetime

## 10.2    DETERMINING THE SIZE REQUIRED

Once again we turn to an example language to explore the issues; in this case, C. In C there are three basic classes of data: primitive data types defined by the compiler, arrays, and structures. C provides a function `sizeof(type)` that returns the size in bytes of its operand. Whether the result of `sizeof` is `unsigned int` or `unsigned long` is implementation defined. The `stdlib.h` header file and `sizeof` are defined consistently in order to declare `malloc` correctly.

## CASE 52. DESIGN AN ALGORITHM TO CALCULATE SIZE

The purpose of this case is to develop an understanding of how `sizeof` is defined. As a hint, `sizeof` is defined by structural induction on the type construction expressions in C.

Before reading on, define your version of `sizeof` using structural induction on the type expressions.

### 10.2.1 Compiler-Defined Primitive Data

Structural induction is recursion. The base cases for the structural induction must be the primitive data types. Use the program shown in Figure 10.2 to explore the values for a computer at your disposal.

### 10.2.2 Space for Structures

Although arrays came first historically, determining an algorithm for space should consider structures next. For this chapter, *structures* are data structures that have heterogeneous data components. In C, structures are `structs` and `unions`, whereas in C++ and Java we have `structs`, `unions`, and `classes`. Object-oriented languages raise another issue: how to deal with *inheritance*; we'll not address that here.

Turning again to Figure 10.2, consider the answer given for `struct a`: 16 bytes. How can that be?

## CASE 53. WHY DOES `SIZEOF` COMPUTE 16 AND NOT 13?

Using test cases you develop, write an explanation as to why `struct a` is 16 bytes and not 13. Be very cautious—remember, `malloc` is involved as well.

Once you have satisfied yourself that you have the correct size and alignment information, define a recursive function that computes the size of structures.

### 10.2.3 Arrays

Arrays are comprised of elements. Arrays differ from structures in that all elements of an array are the same size. This requirement is one of the

```
int prog(int x) {
printf"%d"}xint prog(int y)   {
printf("%d"}y);
struct a {
  char c;
  int i;
  double d;
};

union b {
  char c;
  int i;
  double d;
};

int main(int argc, char* argv[] ) {
 printf( "sizeof(char)\t%d\n", sizeof(char));
 printf( "sizeof(int)\t%d\n", sizeof(int));
 printf( "sizeof(int*)\t%d\n", sizeof(int*));
 printf( "sizeof(double)\t%d\n", sizeof(double));
 printf( "sizeof(struct a)\t%d\n", sizeof(struct a));
 printf( "sizeof(union b)\t%d\n", sizeof(union b));
 printf( "sizeof(union b [5])\t%d\n",
                         sizeof(union b [5]));
 printf( "sizeof(union b [5][6])\t%d\n",
                         sizeof(union b [5][6]));
}
```

**Figure 10.2   Test Program in C for `sizeof`**

justifications for the concept of *variant records* or *unions*. Variant records can have multiple internal organizations but the size of the variant record is fixed.

The design issues with arrays is developing an algorithm to find an arbitrary element of the array. The basic approach is to develop an *addressing polynomial*. The basic model is that the beginning address of an element is computed from some known point, or *base*, plus an *offset* to the beginning of the element. Because all elements are the same size, we have all the information we need.

### 10.2.3.1 One-Dimensional Array Addressing Polynomials

For a one-dimensional array, the addressing polynomial is easy. Consider the C declaration

```
int array[27];
```

What is the address of the thirteenth element? It is instructive to "count on our fingers," because this is the basis of the more complicated situations.

| Element Number | Offset |
|---|---|
| First | 0 |
| Second | sizeof(int) |
| Third | 2*sizeof(int) |
| ... | ... |
| Thirteenth | 12*sizeof(int) |

The language design issue that this illustrates is the question, "Is the first element numbered '0' or '1'?" Mathematical practice (Fortran, for example) specifies that the first element is 1; C practice is that the first element is zero. Why did C developers choose zero? Speed. The upshot of this is that there are two possible addressing polynomials:

| Label | Formula |
|---|---|
| Zero-origin | AP(base, number, size) = base + n*size |
| Unit-origin | AP(base, number, size) = base + (n-1)*size |

### 10.2.3.2 One-Dimensional Arrays with Nonstandard Origins

Many problems are naturally coded using arrays that do not start at their "natural" origin of zero or one. As an example, consider this simple problem:

> Read in a list of pairs of numbers, where the first number is a year between 1990 and 2003 and the second is a dollar amount. Compute the average dollar amounts for each year.

The natural data structure would be a floating point vector (one-dimensional array) with the origin at 1990 and the last index at 2003. Following conventions used by some languages, we would note this *range* as 1990..2003.

## CASE 54. DESIGN AN ADDRESSING POLYNOMIAL FOR ARRAYS

Derive an addressing polynomial expression for the arbitrary case where the lowest element is numbered $l$ and the highest numbered element is $h$.

### 10.2.4 Multidimensional Arrays

Many problems use higher (than one) dimensional arrays. The general case has dimensions that are denoted $m..n$ and each dimension could have its own lower and upper bounds; for the $i$th dimension, denote these bounds as $l_i$ and $h_i$.

Even at the two-dimensional case there is a choice: there are two ways to linearize the four elements of a $2 \times 2$ array. The critical insight comes from writing out the two possibilities.

## CASE 55. DEVELOP AN EXTENDED ADDRESSING POLYNOMIAL

Complete the following exercise:

1. Draw a $2 \times 2$ matrix as a square array.
2. Label each element of the array with the index pair.
3. Linearize in two different ways. **Hint:** The first and last must be the same for each ordering.

How should we interpret these drawings? If it is not clear, construct other examples. Ultimately you should convince yourself that in one case the row index changes more slowly and in the other the column index changes more slowly. We call the first the *row major ordering* and the second the *column major ordering*. Row major ordering is more common but Fortran is column major.

## CASE 56. GENERAL TWO-DIMENSIONAL ARRAY ADDRESSING POLYNOMIAL

Develop the addressing polynomial for the general case of two-dimensional arrays.

## 10.3   RUNTIME STORAGE MANAGEMENT

Runtime storage management implements scoping, allocating, and deallocating local storage. The three concepts are tightly coupled.

### 10.3.1 Scoping as a Visibility Function

The fundamental concept in scoping is to answer the question, "When can a data object be created, retrieved, updated, and destroyed?" The λ-calculus provides the prototypical rules for scoping in programming languages (not to mention logic and other areas of mathematics). The so-called *α-conversion rule* states that the name of bound variables is not important; that as long as I consistently rewrite the names, I will get the same result when I evaluate an expression.

An easy example is the C programs shown in Figures 10.2 and 10.3; when compiled and run, these two provide identical results. As trivial as this may seem, it is a fundamental concept in compiling functions with local variables. Consider now the C program in Figure 10.3. Although this may not seem like a program someone would write, it demonstrates a point: the y on lines (5) to (9) is different from the y on lines (1) to (4) and (10). How should we model this? It is also clear that the same variable z is used in both sections of code. How do we model this?

We will use a visibility function to model the behavior. The visibility function considers the runtime versus definitions. You can think of this as a graph with time on the abscissa and variable names on the ordinate axes (see Figure 10.4).

### 10.3.2 Stack Frame

A *stack frame* or *activation record* is a data structure used to create temporary storage for data and saved state in functions. The word *stack* gives away the secret: the stack frames are pushed onto the *activation stack*

```c
int main(void) {
(1)    int y;
(2)    int z = 10;
(3)    for( y=0; y<=5; y++)
(4)      printf("%d ",y+z);
(5)      {
(6)      double y;
(7)      for( y=0; y<=5; y++)
(8)        printf("%f ",y+z);
(9)      }
(10)   printf("%d\n",y+z);
}
```

**Figure 10.3   Simple C Program Illustrating Visibility**

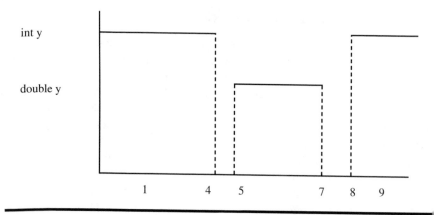

**Figure 10.4    Illustration of Visibility (The graph shows statement execution on the abscissa and visibility on the ordinate axis. Solid lines indicate when a variable is visible.)**

(hence the synonym *activation record*). The term *stack frame* is somewhat misleading in one sense: that the entire frame exists on the stack all at once; this is not necessarily so because this can waste stack space.

## CASE 57. STACK FRAME MANIPULATION

Develop a function vocabulary to manipulate stack frames and the activation stack. This vocabulary is necessary for compiling functions and local variables to functions.

The stack frame will have space for all user-defined local variables, but there may be many other frame slots used by the compiler to save such things as temporary variables generated during expression compilation and administrative information.

From the example program, Figure 10.3, it is clear that the activation stack cannot be a *true* stack because we can see into stack frames other than the top. This indicates a list structure with a stack structure imposed. The performance issue is that we cannot afford to search for the correct stack frame.

## CASE 58. LISTING ACTIVATION RECORDS

Expand the vocabulary developed above to incorporate the list requirement.

# 11

d hence need
he software
agement
s to be

# ..L SOFTWARE
# DESIGN PROCESS PORTFOLIO

---

**Reference Materials**

Watts Humphrey. *Introduction to the Personal Software Process.* Addison Wesley Professional. 1996.

**Online Resources**

www.gnu.org: GNU Foundation. *GForth Reference Manual*

Software Engineering Institute: http://www.sei.cmu.edu/tsp/psp.html

PSP Resources Page developed by the Software Engineering Group at the University of Karlsruhe: http://www.ipd.uka.de/mitarbeiter/muellerm/PSP/

---

The purpose of the Personal Design and Software Process Portfolio (PDSPP) portion of the course is to introduce professional concepts of personal time management and continuous improvement. My experience is that students (and faculty, including me) often have poor time management skills. This is due partly to difficulties in budgeting time and partly to the nature of research. Regardless, we all can benefit from occasionally taking strict accounting of our time to discover ways we can improve.

There is a very practical aspect to this exercise. This text is meant for seniors in computer science who will soon be looking for a job. I have had many students tell me that showing recruiters a portfolio drawn from their classes is an extremely effective tool in job interviews. Professionally, a complete portfolio of a year's work is a powerful message at annual review time.

The outline of a portfolio is described in this chapter. The goal is to present practical advice on developing yourself as a professional in computer science. Many of the ideas for this chapter come from the concepts put forth in Humphrey's *Personal Software Process* or PSP as it is generally referred to (Humphrey 1996). However, PSP does not address issues relating to system design. My experience has been that undergraduate students rarely have experience in developing an entire system an̶ help in this area. For this reason, the text must address both t̶ process and the design process. This chapter develops time ma̶ portions of PDSP. Most milestones will have unique design concep̶ practiced in the milestone.

## 11.1 TIME MANAGEMENT FOR COMPUTER SCIENCE STUDENTS

It is a simple fact that there are not twenty-four usable hours in a day, that a three-credit university course is more like three hours a week, and you can't sit down Thursday night to write a program due Friday morning.

### 11.1.1 There Are Not Twenty-Four Hours in a Day

Although the clock says there are twenty-four hours in a day, you cannot work twenty-four hours a day. For health's sake, you need seven to eight hours of sleep a day and about three hours for personal hygiene, meals, etc. It really looks like you can only have fourteen to fifteen hours per day to spend as you want. In a perfect world, you would have $7 \times 15 = 105$ hours to spend as you like.

Anyone who has been associated with college students knows that students do not spend Friday night, Saturday, or most of Sunday doing school work. And that is as it should be. However, this effectively removes thirty more hours from your time budget: you now have $5 \times 15 = 75$ hours per week to allocate.

### 11.1.2 A Three-Credit Course Is Nine Hours

A three-credit university course is really nine hours because you are expected to spend three hours in the classroom and two hours per hour of class preparing for class. The preparation may be reading or homework to hand in or it may be working on a program. Therefore, a fifteen-hour load requires that you spend forty-five hours per week. Although the "standard" work week is forty hours, you can expect to spend a bit more than that each week. Notice, though, that you have seventy-five disposable hours per week.

Unfortunately, most of us cannot be so efficient as to use all the time in-between class. So, let's say that you spend your days in class (fifteen hours) and that leaves thirty to forty-five hours at night; that's six to seven hours per night. My experience is that you cannot profitably do that, either. After a long hard day, you are not an efficient user of time. The suggestion is that you find a block of time during some days to get the nighttime load down to something manageable, say four hours a night.

| Notation | Explanation |
|----------|-------------|
| nnn | Class number |
| CP(nnn) | Preparing for Class nnn |
| C(nnn) | Attending Class nnn |
| D(nnn) | Designing programs for Class nnn |
| CT(nnn) | Coding and testing for Class nnn |
| U(nnn) | Documenting programs for Class nnn |

**Figure 11.1 Notations for Documenting TUL Form**

### 11.1.3 The Key Is Finding Out How You Spend Your Time

The key to any science is the metrics used to measure activities. The Time Utilization Log (TUL)* is to be used to record how you're using your time. It has the class times† already accounted for between 8:00 a.m. and 5:00 p.m. The rest of the day is blocked off in half-hour increments. The assumption is that you stay up late and get up late. Early risers should modify the grid.

## CASE 59. TRACKING YOUR TIME USAGE

You are to insert your schedule onto the spreadsheet using the codings shown in Figure 11.1. On the right side is a place to list major activities that you typically have during the week that are not accounted for by the set notation with a code of your own. I would suggest that you code activities that can take up to a half-hour.

During the week, keep track of your time and activities and enter them into the log. If you skip a class, be honest, and remove it from your schedule for that week. However, skipping class is a slippery slope to disaster. In my classes, you are allowed to skip a maximum of the number of hours for the course. After that, it's an automatic F.

### 11.1.4 Keeping Track of Programming Time

The Time Recording Template (TRT) (note that this form is one of many in the Excel packet) is to be used for recording time spent "programming," which includes time to design, code, test, and document programs. As with

---

\* This is not a standard part of PSP. However, I find most students cannot account for their time and this is a good place to start.

† Clemson has two schedules: Monday–Wednesday–Friday and Tuesday–Thursday. Alter this schedule to fit the local situation.

the TUL, the TRT keeps track of your time but the TRT is used to determine how you use your time when developing programs. You should have an entry for every time you program for this class. The columns should be self-explanatory except the phase column. The phase designations are listed on the right of the sheet. You may add your own phase codes and explanations if you think them necessary.

## 11.2 DISCOVERING HOW LONG IT TAKES TO WRITE A PROGRAM

It is an unfortunate truth that developing programs that are rigorously tested is a time-consuming business. Experiments in my classes have validated a long-held folk theorem: programmers produce approximately 2.5 debugged lines of code per hour. The best I have seen in my classes is right at 3 lines per hour. In this course, you may well have several milestones that are 100 to 120 lines of code. At 2.5 lines of code per hour, you will spend approximately 40 to 48 hours to complete such a milestone. Even if given 2 weeks to complete the milestone, you need 20 to 24 hours per week.

In order to properly plan for your work, you must know how many lines of code a particular project requires *before* you start. How can we estimate how many lines of code a problem may take?

To begin with, what constitutes a line of code (LOC)? Certainly not blank lines and comments. Clearly, there are two categories of lines: data declaration and algorithmic. Fortunately, data declarations don't change much once they're written; however, they do take time. For very simple declarations, the time is negligible. You must learn to judge where the breakpoint is between *simple* and *complicated* data declarations.

Algorithms are a different problem entirely. The problem is that most real-life algorithms are very long and complex (see the 100- to 120-line milestone quoted above). Any long and involved algorithm must be decomposed into smaller blocks of code until one can accurately "guesstimate" how many lines are needed. An organized way to analyze an algorithm is to produce a work breakdown structure (WBS) (Figure 11.2).

| Task Number | Description | Inputs From | Outputs To | Planned Value | Earned Value | Hours | Cum. Hours | Date |
|---|---|---|---|---|---|---|---|---|

**Figure 11.2   Work Breakdown Structure Headings**

## 11.2.1 Problem Decomposition

Before explaining the WBS, we should first explore how one goes about decomposing a problem. This process relies on your having gained a certain level of expertise: you should be able to determine what parts of the problem you understand how to accomplish and what you must work harder on. Here's an example:

> Write a program in your favorite language that reads in a series of floating point numbers, one to a line, and determines the minimum and the maximum. At the end of file, print out the minimum first and the maximum second.

Below is the transcript of a session that decomposes the above problem into components:

> This is a full program, so there are three phases: input, process, and output. The process part is a while loop based on the "until end-of-file" clause. There's no range specified, so I have to initialize the two variables, call them min and max, but I don't have any particular value. OK, then read in the first one ... Oops, note that there might not be a first one ... init with the first one. Now you can have a loop with three statements: (1) check min changes, (2) check max changes, and (3) read next one.
>
> Now the conditions on the while loop: need to have a variable—call it num_read—set by the input statement. Use num_read not zero to continue.
>
> After the while loop, print out the two values. Oops, almost forgot the null case. Hm ... there's no spec on what to do there; print out nothing, I guess.

The approach given here is technically called *talking a solution* and this technique is very useful in decomposing a problem. You can use this technique yourself by either writing out your thoughts or by using a voice-activated tape recorder.

## 11.2.2 Work Breakdown Structure

The WBS is just a form that aids the decomposition process. A logical way to begin is to decompose the problem and identify the components. It may be more natural to have a graphic representation of the components and their couplings (connections). Number each component and enter this as the "Task Number" in the form. Enter the component name and description. Fill in the "Inputs From" and the "Outputs To" columns based on the task number of the input and output components. Remember that at this level inputs may come from many different components and the outputs may go to many components. For all but the simplest projects, a graphical representation of this information is a must. We are now ready to estimate the effort.

Now estimate the number of "Lines" per component. The estimated "Hours" should be computed by either your measured rate or the 2.5 lines per hour rate that is the historical average. Now sum the "Hours" column to produce the cumulative hours, "Cum. Hours."

We have now completed the planning portion. However, the WBS must be used during development to determine whether or not you can complete the project on time. In the next section we discuss *earned value*.

## 11.3 EARNED VALUE COMPUTATIONS

The WBS and scheduling planning template have a column for cumulative hours under the heading of "Earned Value" which stands for *cumulative earned value*. Earned value is a concept of project management used in many large companies and by all government contractors (Fleming and Koppelman 2000). The basic idea is that every project has a value but that value is not just a percent completion of a particular component because it is almost impossible to reliably estimate percent complete. Earned value is about costs of resources (time, people, equipment). For class, we will equate planned hours as planned earned value and actual hours as actual costs because that time is the only capital you have to invest.

### 11.3.1 Baseline and Schedule Variance Computations

For illustration, suppose we have six modules *A* to *F* that must be delivered in two weeks. Suppose we estimate the number of time units (say hours) that this task should take as 100 and the amount of time needed for each as shown in Figure 11.3, called the baseline. In the baseline, the tasks are presented with the planned time estimates.

As the programming work is performed, value is "earned" on the same basis as it was planned, in hours. Planned value compared with earned value measures the volume of work planned versus the equivalent volume of work accomplished. Any difference is called a *schedule variance*. In contrast to what was planned, Figure 11.4 shows that work unit *D* was not completed and work unit *F* was never started, or 35 hours of the planned

|         | A  | B  | C  | D  | E  | F  | Total |
|---------|----|----|----|----|----|----|-------|
| Planned | 10 | 15 | 10 | 25 | 20 | 20 | 100   |

**Figure 11.3  Planned Workload**

| | A | B | C | D | E | F | T otal |
|---|---|---|---|---|---|---|---|
| Planned | 10 | 15 | 10 | 25 | 20 | 20 | 100 |
| Earned | 10 | 15 | 10 | 10 | 20 | 0 | 65 |
| Schedule | 0 | 0 | 0 | −15 | 0 | −20 | −35 |

**Figure 11.4   Earned Value Computation (The time invested in the various modules is compared with the planned time and the variance is computed. The goal is to have no variance.)**

work was not accomplished. As a result, the schedule variance shows that 35 percent of the work planned for this period was not done.

Schedule variance really only makes sense if applied relative to completion when we also know exactly what must be done.

### 11.3.2   Rebudgeting

In this example, we're really not in too bad shape. Four modules are complete and we estimated them correctly but we have badly miscalculated how long it takes to do the other two (D and F). The work still has to be done, but the next milestone looms large. The only thing you can do at this point is to consider merging the effort for D and F with what has to happen in the next milestone.

There are many lessons here:

- If you cannot correctly estimate your effort and consistently provide fewer than necessary resources, you're doomed.
- Overestimation is just as bad as underestimation. From overestimation, you waste resources that could be used to do other things.
- You must keep track of how much resources you've expended in order to not over- or undercommit.
- Accurate estimation is the key to this whole process.

### 11.3.3   Process

The process of developing an earned value plan is exactly the process we need to design a program. The most important step is the WBS. Every milestone will start with developing a WBS, which includes everything that is required to successfully complete the project. The key to success is to account for everything! For example, the following classes of work need to be accounted for:

- Class time
- Class preparation time

- Planning time
- Design time
- Programming time
- Test development time
- Debugging time

Note to the wise: Do not forget the effort for your other classes and extracurricular activities.

Because the design and implementation of each subprogram method and class structure take time, those times show on the WBS as well. If you are honest with yourself and honestly estimate your required effort, you will quickly see that you must be very disciplined. There is a sobering statistic: over the past 50 years, an accomplished programmer has been able to produce only 2.5 debugged statements per hour and this figure has been validated in several of the author's classes.

As a procedure to develop each milestone:

1. When we start a new milestone, develop a preliminary WBS that defines each task. Large tasks are broken into manageable work packages using task decomposition techniques. The WBS normally follows this hierarchy: the program that represents the solution to the milestone is broken into data structures (class variables) and algorithms (methods). (There is a well-defined OOPS design methodology: make use of it!) The WBS must contain, at a minimum, an entry for every method because every method takes time to write. This WBS will be its most accurate if it is based upon the very short, well-defined tasks.

2. The instructor determines the overall development budget based upon the instructor's estimate of the work. This will normally be given as a number of lines of code (LOC). I generally estimate code based on how many pseudo-code algorithm lines will be required. I tend not to count type statements unless there are a significant number of structure type statements. Estimating LOC is an art, not a science, and I shoot for an order of magnitude, say, to the nearest ten lines. $LOC/2.5$ is the number of hours.

3. You then will give an initial budget based upon the instructor's estimate. In the past, whenever the question of extension has come up, I generally made that decision based on my perception of how close the better students are. With this earned value discipline, I will have objective measurements.

4. The WBS is now the basic implementation plan.

5. As the milestone proceeds, you must update the plan. Based upon the more current estimates, time shifts may be possible only if reasonably documented and a general pattern of slippage is evident.

6. A new WBS/budget sheet with any variances will be due each Friday. Using the WBS/budget sheet as the input, a spreadsheet can be run by the instructor. The weekly WBS updates should be more detailed as design and implementation proceed.
7. Scheduling information is discussed each Friday in class.
8. Students should come prepared to explain their WBS status.
9. Plans are adjusted and actions taken to address any problems that are indicated by the reports.
10. You should assign planned start dates and completion dates to each subtask in accordance with the milestone schedule.
11. The final WBS will be handed in with the milestone reports.

### 11.3.4 Templates

The Personal Design and Software Process (PDSPP) documentation templates are provided as an Excel workbook. This section provides a key to that workbook based on the tab names in the Excel workbook. The original Excel workbook was retrieved online from the *PSP Resources* page at the University of Karlsruhe, Germany.* There are many such sites on the Internet. The *Time Utilization Log* was developed for the Clemson University time schedules.

The Karlsruhe packet contains all the PSP forms mentioned in Humphrey's book. The instructor should choose how much of the PSP discipline he or she is willing to impose. **Note:** Every form has a header that asks for what should be obvious information. For example, PSP0 asks for your name, due date, milestone, instructor, and language of implementation; this information should be obvious.

## 11.4  SUMMARY INFORMATION FOR PDSPP0

PDSPP0 forms are only used during the implementation phase of the course; otherwise, such fields as *milestone number* are meaningless.

■ Header. This should include your name, due date, milestone, instructor, and language of implementation. Fill these out with the obvious.
■ Phases. There are some standard phases that are always in play, such as planning and design, and should always be on the form. Add phases ("time sinks") for your unique situation. There are approximately 16 hours (960 minutes) of "usable" time during the day and approximately 180 minutes for meals. Therefore, you should

---

* http://www.ipd.uka.de/mitarbeiter/muellerm/PSP/#forms

account for about 750 minutes per day (3750 minutes per five-day week). Don't forget recreation and other obligations, such as civic activities. The sense I get from talking to students is that a 50- to 60-hour work week is not uncommon in computer science. Enter the planned and actual time in each phase. Planned values are estimates so they may well be a changing number as you understand more about the project.

- Enter the planned time in minutes you plan to spend in each phase.
- Enter the actual time in minutes you spent in each phase.
- Enter your best estimate of the total time the development will take.
- Time — To Date. Enter the total time spent in each phase to date. Since we're always working on a milestone, there is always something to charge time to.
- Time — To Date %. Enter the percent of the total "To Date" time that was spent in each phase. This form is a summary form. You can use Excel to compute these times for you. Your actual time usage should be consistent with the Time Recording log (TRT).
- Defects. Defects are not planned, so there is no way you can budget time for them, but you can count them and analyze why you are injecting them. This information comes from the Defect Log (DLT)
- Defects Injected and Defects Removed. Enter the actual numbers of defects injected and removed in each phase. Careful planning of Excel will simplify these calculations.
- Defects — To Date. Enter the total defects injected and removed in each phase to date.
- Defects — To Date %. Enter the percent of the total "To Date" defects injected and removed in each phase.

## 11.5   REQUIREMENTS TEMPLATE (RWT)

- Header. Enter the name, date, instructor, and milestone as on the summary.
- Item. Identification; could be a number or mnemonic.
- Inputs. What are the inputs to this constraint or criteria?
- Criteria or Constraints. Criteria are logical statements about what the system is supposed to do. Constraints are performance standards. Requirements are the first full step in the Dewey Problem Solving Paradigm.

## 11.6 TIME RECORDING LOG (TRT)

- Header. Name, date, instructor, and program number as in the summary.
- Date. Enter the current date.
- Start. Enter the time in minutes when you start a project phase.
- Stop. Enter the time in minutes when you stop work on a project phase, even if you are not done with that phase.
- Delta time. Enter the elapsed start to stop time less the interruption time.
- Phase. Note the phase on which you were working by using the phase name.
- Comments. Describe the interruption to the task you were doing and anything else that significantly affects your work.

## 11.7 SCHEDULE PLANNING TEMPLATE (SPT)

- Header. Enter the name, date, instructor, and milestone as on the summary.
- Number (No). This is the number of the work period.
- Date. Date the work period took place.
- Plan. How many hours do you plan to work this period and what are the total cumulative hours?
- Actual. Hours and Cumulative Hours as above. Cumulative EV and Adjusted EV are described below.

## 11.8 DEFECT LOG TEMPLATE (DLT)

A defect is anything in the program that must be changed for it to be properly developed, enhanced, or used.

- Header. Enter the name, date, instructor, and milestone as on the summary.
- Date. Enter the date when you found and fixed the defect.
- Number. Enter a unique number for this defect. Start each project with 1.
- Type. Enter the defect type from the defect type standard. See tab DCL.
- Inject. Enter the phase during which you judge the defect was injected.
- Remove. Enter the phase in which you found and fixed the defect.
- Fix time. Enter the time you took to fix the defect. You may time it exactly or use your best judgment.

■ Fix defect. If this defect was injected while fixing another defect, enter the number of that defect or an X if you do not know.

The Defect Codes Legend (DCL) is a table of possible causes and defects. You may add to your own personal copy any defect information that will help you spot defect causes in your process.

| Cause Code | Defect Type | Description |
|---|---|---|
| | 10 | Documentation comments, messages |
| ed (education) | 20 | Syntax spelling, punctuations, formats |
| co (communication) | 30 | Build, package change management, library, version control |
| ov (oversight) | 40 | Assignment declaration, duplicate names, scope, limits |
| tr (transcription) | 50 | Interface procedure calls and references, I/O, user formats |
| pr (process) | 60 | Checking error messages, inadequate checks |
| | 70 | Data structure, content |
| | 80 | Function login, pointers, loops, recursion, computation |
| | 90 | System configuration, timing, memory |
| | 100 | Environment design, compile, test, other support system problems |

# 12

---

# HOW DID WE GET HERE? WHERE DO WE GO FROM HERE?

Why is the chapter on history at the end of the text? Shouldn't it be first? Not necessarily. The purpose of this chapter is to consider "the past as prologue" and "possible histories from this point on."

It would be tempting to think that the current state of the art in programming languages is the pinnacle of development. It would be tempting to believe that the object paradigm is the last improvement in programming language design. Technology has a way of making a mockery of such thoughts. The purpose of this chapter is to consider "What's past is prologue" (William Shakespeare, *The Tempest*).

This chapter models how historical studies are conducted. You might call this the "follow the string" approach. I will choose some ends of the string by making observations about the state of programming languages and follow these strings backward. If we were going to write a paper on the subject, we would then rewrite the timeline forward. There is a danger in this forward exposition because the reader never quite knows where the author is going or why a certain point might be germane.

## 12.1 LOOKING BEHIND

In order to make sense of the roots of computing (which is, after all, the reason for programming) we start with the first programming language and work backward. We start with the question, "Why was FORTRAN even invented?" We can actually trace it back to the Hindus in 2000 BCE.

In choosing historical material, it is tempting to take 1954 as the beginning of the Universe; November 1954 is given as the date of the introduction of FORTRAN. However, this makes little sense, as FORTRAN, which stands for *Formula Translator*, did not spring *de novo* from out of the ether. FORTRAN was put in place to give scientists a mechanism to implement numerical routines on computers. From our investigation of linguistics, we can see that *pragmatics* was the driver of language development. The form of FORTRAN is not surprising to anyone who has studied numerical analysis texts of the 1950s and 1960s.

### 12.1.1  The Difference Engine

Tracking this "scientific programming" heritage backward, we next find Ada Byron. In the late 1970s, the programming language Ada was all the rage. The foundation of the pragmatics of Ada was that of embedded systems, such as guidance systems on missiles. The reason to bring Ada up is the historical note that Ada was named for Ada Byron, Countess of Lovelace, and the daughter of the poet Lord Byron. Lady Byron was a very complex person who teamed with Charles Babbage, who invented the Difference Engine. The purpose of the Difference Engine was to solve differential equations for commerce. Today's computers are legitimate heirs of the Difference Engine, which was successfully implemented at Manchester University in 1948 by the engineers F. C. Williams and Tom Kilburn.

### 12.1.2  The Golden Age of Mathematics

Some of the issues that confronted Babbage and Lovelace were first explored by Isaac Newton and his contemporaries. There are so many famous contemporaries of Newton that it is impossible to list them all. For our search, though, we are interested in two: Colin Maclaurin and Brook Taylor. Newton, Maclaurin, and Taylor developed the fundamental principles for deriving polynomials for computing functions. The simple realization that infinitely presented numbers ($\pi$, for instance) must be finitely presented requires us to redevelop continuous mathematics with discrete mathematics. In particular, the computation world requires algebraic, not analytic (calculus), representations. Another important link at this juncture is the Bernoulli family.

Viète began the use of symbols in his book *In artem analyticam isagoge* in 1591.*

### 12.1.3  Arab Mathematics

The history of algebra is what we want to track, apparently. Most computer scientists should know that the word "algorithm" is of Arabic origin; in fact, so is the word "algebra." *Algebra* comes from the title of

---

* http://www-groups.dcs.st-and.ac.uk/ history/Mathematicians/Viete.html

a book on the subject, *Hisab al-jabr w'al muqabala*,* written about 830 by the astronomer/mathematician Mohammed ibn-Musa al-Khowarizmi. *Algorithm* is a corruption of al-Khowarizmi's name. However, the Arabian study of algebra came late, following the Hindus, Greeks, and Egyptians. The most advanced studies of algebra in antiquity were carried out by the Hindus and Babylonians.

## 12.1.4  Hindu Number Systems and Algebra

Although the Hindu civilization dates back to at least 2000 BCE, their record in mathematics dates from about 800 BCE. Hindu mathematics became significant only after it was influenced by the Greeks. Most Hindu mathematics was motivated by astronomy and astrology. Hindu mathematics is credited with inventing the base ten number system; positional notation system was standard by 600 CE. They treated zero as a number and discussed operations involving this number.

The Hindus introduced negative numbers to represent debts. The first known use is by Brahmagupta about 628. Bhaskara (b. 1114) recognized that a positive number has two square roots. The Hindus also developed correct procedures for operating with irrational numbers.

They made progress in algebra as well as arithmetic. They developed some symbolism that, though not extensive, was enough to classify Hindu algebra as almost symbolic and certainly more so than the syncopated algebra of Diophantus. Only the steps in the solutions of problems were stated; no reasons or proofs accompanied them.

The Hindus recognized that quadratic equations have two roots, and included negative as well as irrational roots. They could not, however, solve all quadratics because they did not recognize square roots of negative numbers as numbers. In indeterminate equations the Hindus advanced beyond Diophantus. Aryabhata (b. 476) obtained whole number solutions to $ax \pm by = c$ by a method equivalent to the modern method. They also considered indeterminate quadratic equations.

The Greek mathematician Diophantus (200 CE? 284 AD?) represents the epitome of a long line of Greek mathematicians (Archimedes, Apollonius, Ptolemy, Heron, Nichomachus) away from geometrical algebra to a treatment that did not depend upon geometry either for motivation or to bolster its logic. He introduced the syncopated style of writing equations, although, as we will mention below, the rhetorical style remained in common use for many more centuries.

Diophantus' claim to fame rests on his *Arithmetica*, in which he gives a treatment of indeterminate equations—usually two or more equations

---

* Often translated as *Restoring and Simplification* or *Transposition and Cancellation.*

in several variables that have an infinite number of rational solutions. Such equations are known today as "Diophantine equations." He had no general methods. Each of the 189 problems in the *Arithmetica* is solved by a different method. He accepted only positive rational roots and ignored all others. When a quadratic equation had two positive rational roots he gave only one as the solution. There was no deductive structure to his work.

### 12.1.5  Greek Geometrical Algebra

The Greeks of the classical period, who did not recognize the existence of irrational numbers, avoided the problem thus created by representing quantities as geometrical magnitudes. Various algebraic identities and constructions equivalent to the solution of quadratic equations were expressed and proven in geometric form. In content there was little beyond what the Babylonians had done, and because of its form geometrical algebra was of little practical value. This approach retarded progress in algebra for several centuries. The significant achievement was in applying deductive reasoning and describing general procedures.

### 12.1.6  Egyptian Algebra

Much of our knowledge of ancient Egyptian mathematics is based on the Rhind Papyrus. This was probably written about 1650 BCE and is thought to represent the state of Egyptian mathematics of about 1850 BCE. Egyptian mathematicians could solve problems equivalent to a linear equation in one unknown. Their method was what is now called the "method of false position," which is also a modern numerical algebra technique. Their algebra was rhetorical, that is, it used no symbols; problems were stated and solved in full language.

The Cairo Papyrus of about 300 BCE indicates that by this time the Egyptians could solve some problems equivalent to a system of two second-degree equations in two unknowns. Egyptian algebra was undoubtedly retarded by their cumbersome method of handling fractions. Egyptian fractions are a favorite programming project in beginning programming projects and data structures classes.

### 12.1.7  Babylonian Algebra

The mathematics of the Old Babylonian Period (1800–1600 BCE) had a sexagesimal (base 60) system of notation that led to a highly developed algebra system. The Babylonians had worked out the equivalent of our quadratic formula. They also dealt with the equivalent of systems of two equations in two unknowns. They considered some problems involving more than two unknowns and a few equivalent to solving equations of higher degree.

Like the Egyptians, their algebra was essentially rhetorical, meaning that there were few symbols and that problems and procedures were described in natural language. The procedures used to solve problems were taught through examples and no reasons or explanations were given. Also like the Egyptians they recognized only positive rational numbers, although they did find approximate solutions to problems which had no exact rational solution.

### 12.1.8  What We Can Learn

The very nature of algebra and algebraic notation changed over its 40-century history. From approximately 1800 BCE to around 200 BCE, algebra was primarily rhetorical, meaning that it was written in natural language. From 200 BCE to 1500 CE, algebra was syncopated, meaning that it was comprised of both natural language and symbols. The modern *symbolic form* appeared in the sixteenth century. The modern *abstract* version appeared only in the nineteenth century.

Algebra was driven by linguistically pragmatic issues. For example, the "lowly" quadratic formula appeared early but generalized polynomial notation arrived much later, requiring the symbolic version of algebra to appear.

The general theory of the algebra of real numbers began to appear in its modern form in the mid-1850s. Modern algebraic studies were greatly expanded by the use of algebraic methods with logic. Modern algebraic approaches use *category theory*, an approach that emphasizes functions over conditions. Category theory is heavily used in theoretical work in computer science.

## 12.2  THE ROLE OF THE λ-CALCULUS

We have called the definition of how a program language works its semantics. There have been three approaches to defining semantics: (1) operational, (2) denotational, and (3) axiomatic or connotational. Operational semantics approaches posit a hypothetical machine that is then used to describe the operation of a program. Denotational semantics is based on λ-calculus, and axiomatic semantics approaches are based on logic.

In 1963, John McCarthy (the "father" of Lisp) wrote an article that would change the way computation was perceived. Until McCarthy's article, computation, and by implication programming and programming languages, was tied to Turing machines. Even today, Turing machines (not λ-calculus) are the basis of algorithms research in the *NP*-complete investigations. McCarthy's article instead proposed λ-calculus as the foundation (McCarthy 1963).

In 1956, Noam Chomsky developed what is now called the *Chomsky Hierarchy*, which showed how Turing machines and grammars were connected (Chomsky 1956, 1959). In effect, Turing machines were shown to be equivalent to the so-called Type-0 Grammars. Chomsky's work tied together three major languages for computation: the λ-calculus, Turing machines, and grammars/languages. Turing machines and λ-calculus were already linked.

Turing machines were introduced by Turing in 1936 in a paper, "On Computable Numbers, with an Application to the *Entscheidungsproblem*." In the paper, Turing tried to captured a notion of algorithm with the introduction of Turing machines. The *Entscheidungsproblem* is the German name for Diophantine problems. Turing showed that the general Diophantine equations could not be solved by algorithms.

A few months earlier Alonzo Church had proven a similar result in "A Note on the Entscheidungsproblem" but he used the notions of recursive functions and lambda-definable functions to formally describe what the logicians call *effective computability* (Church 1936). Lambda-definable functions were introduced by Alonzo Church and Stephen Kleene (1936), and recursive functions by Kurt Gödel and Jacques Herbrand. These two formalisms describe the same set of functions, as was shown in the case of functions of positive integers by Church and Kleene (Church 1936, Kleene 1936).

The upshot of Church, Gödel, Kleene, and Turing was that although there were many different languages to describe algorithms, they were equivalent, in the sense that any computable function coded in one language could be coded in any of the other languages.

It is important to note that the issue of computation in the 1920s and 1930s was an issue in logic, not computing as we know it today. The study of equivalence of algorithms is an issue in the theory of computation or computability.

The λ-calculus grew out of Church's earlier work published in 1927. Church was interested in a single question: how to formally describe how variables work. In the 400 plus years since Viète, the actual, precise rules on the use of variables and parameters had not been formalized; the λ-calculus provided such a basis. Interestingly, the actual semantics for the λ-calculus did not get worked out until the 1980s.

In summary, the role of functionality in programming is undeniable. The notation used in programming languages borrows heavily from the standard algebraic notation. This leads to a class of languages, called functional languages, that include Lisp and ML. ML's pragmatic beginnings were support of an extensive type system and a demonstration that certain principles of typing could be efficiently implemented. Functional languages are developed on a different basis than the C-based systems, the latter being called imperative to emphasize the use of commands such as `while`. Functional language aficionados maintain that recursion *pragmatically* better captures `whiles`.

## 12.2.1 The Influence of Logic

Modern logic systems are effectively symbolic languages with no specific meaning assigned to the nonlogical symbols. Meaning (semantics) is supplied by assigning variables, constants, and functions a specific meaning based on the use at hand. For example, mathematical theories, such as group theory, outline the logical properties of groups, but leave the specific actions to the semantic interpretation; hence the role of semantics in programming languages.

Another use of logic in programming languages is the development of axiomatic semantics, proposed primarily by Floyd, Hoare, Gries, and Dijkstra. Let's start with Hoare's "An Axiomatic Basis for Computer Programming" (1969). Operational and functional approaches describe how something is computed; the axiomatic approach is to consider under what conditions something is defined. We now call the notational concept *Hoare triples*, denoted:

{*pre-condition*} {*statement*} {*post-condition*}

where *pre-conditions* and *post-conditions* are logical statements about the *state* of the program.

These ideas are the foundation of formal methods, such as Communicating Sequential Processes and program synthesis methods. Students will recognize the software engineering terms *pre-* and *post*-conditions that got their start in Hoare logics.

To summarize, on the surface, one would think there is no connection between logic and programming languages. Clearly this is not true. In fact, it is hard to separate modern logic and modern functional approaches at the foundation level—they all grew out of the larger issue of how to specify logical semantics.

## 12.2.2 Computability

A central issue in computation is the question, "What are the limits of computational devices?" In other words, what sorts of problems are solvable by computation? Turing showed there are some problems that are *inherently unsolvable*. We can see that this problem has dogged mathematics since Diophantus in the third century CE. The focus on computation was an outgrowth of two controversies: Russell's paradox and L. E. J. Brouwer's criticism of mathematics.

**Russell's paradox** was identified after the German logician Gottlob Frege was preparing to publish a book on the newly developed subject of set theory. In the book, Frege allowed for a *set of all sets*. Basically, Russell showed this to be a paradox. There are many types of paradoxes; two are the logical and semantic paradoxes. A logical paradox is a statement that is an argument, based on acceptable premises and using justifiably valid reasoning that leads to a conclusion that is self-contradictory. The second type of paradox is a semantic paradox, which involves language and not logic. A famous semantic paradox is "This sentence is not true"; it is a semantic paradox because the paradox relies on the meaning of the words. Russell's statement is a logical paradox because it is a logical statement (it has premises and reasoning) that Russell showed was oxymoronic. Let $A$ be the set of all sets. Then

(1)  "$A \in A$. $A$ cannot be in $A$ because $A$ is the set of all sets.

(2)  "$A \notin A$. But $A$ is the set of all sets, so it must be in $A$.

(3)  These are the only possibilities.

Russell's paradox set the mathematics and logic world on a hunt for absolute "knowability" in logic. Until this point, it had been assumed that paradoxes were oddities; now the foundations of mathematics were in jeopardy.

## 12.3  MOVING FORWARD FROM 1954

The first recognizable programming language is taken to be first described in 1954, in *Preliminary Report: Specifications for the IBM Mathematical FORmula TRANslating System*. FORTRAN is credited to John Backus and was released commercially in 1957. The specifications of FORTRAN have been standardized since 1966. This standardization is controlled by the J3 subcommittee of the International Committee for Information Technology Standards (INCITS), formerly known as the National Committee for Information Technology Standards (NCITS).

> J3 developed the FORTRAN 66, FORTRAN 77, FORTRAN 90, FORTRAN 95 and FORTRAN 2003 standards. FORTRAN 2003, published 18 November 2004, is an upwardly-compatible extension of FORTRAN 95, adding, among other things, support for exception handling, object-oriented programming, and improved interoperability with the C language. Working closely with ISO/IEC/JTC1/ SC22/WG5, the international FORTRAN standards committee, J3 is the primary development body for FORTRAN 2008. FORTRAN 2008 is planned to be a minor revision of FORTRAN 2003.*

---

* Taken from the J3 Web site,www.j3-FORTRAN.org

Why bring this up? The point is that even such an "ancient" language as FORTRAN is an evolving system, despite the hubris of different programming language communities.

Let's look at the evolution of programming language concepts over the past 50-plus years.

## 12.4  FROM 1954 UNTIL 1965

Programming language development was rampant in the time frame 1954 to 1964. Virtually all the programming language paradigms that survived until today have their basis in this time period.

### 12.4.1  Imperative Languages from 1954 to 1965

Early languages were, not unexpectedly, somewhat ad hoc in their design. The years 1957 to 1965 were very eventful, with the introduction of many of the languages still in use today. Of the imperative style programming language, the three main languages of the period are

1.  FORTRAN: FORTRAN II in 1957 and ending with FORTRAN IV in 1962.
2.  Algol: Algol 58 and Algol 60 in those years; a variant of Algol, called Jovial, was introduced, too.
3.  COBOL: COBOL versions in 1958, 1961, and 1962.

FORTRAN and Algol were primarily conceived as numerical processing languages but rapidly became general purpose languages. COBOL was the product of Rear Admiral Grace Murray Hopper. Whereas Algol and FORTRAN were inspired by scientific computing, COBOL was developed to meet business needs. The difference in language was presentation: although FORTRAN and Algol looked like C-based mathematical languages, COBOL looked like English. FORTRAN and COBOL both have modern, current standards in effect.

Three important offshoots of Algol were Pascal, CPL, and Simula I. Pascal began as an attempt to produce a language that could be used to teach programming; it would become an important commercial success as a system development language. CPL would be extended to BCPL, then B and finally C. Simula I (1964) would become Simula 67, the beginning of the object-oriented paradigm. Simula-based systems were developed not for general programming, but for discrete-event simulation.

Late in this timeframe came another venerable system: Basic in 1964. The reader probably has used either Visual Basic or its modern counterpart VB.NET.

PL/I was first produced in 1964 and was hailed by IBM as its preferred mode of development languages. PL/I was a very rich language that tried to cover the entire landscape of applications. A major difference with PL/I and C-based languages was that the input/output sublanguage was massive, covering access methods that are now subsumed by database management systems. In many ways, C was developed as the antithesis of PL/I.

### 12.4.2  Lisp and the Functional Paradigm

Lisp 1 was produced in 1958 by John McCarthy, then at the Massachusetts Institute of Technology. McCarthy developed Lisp because he disagreed with the approach taken in FORTRAN. Lisp is important for three major reasons: functional programming, introducing the $\lambda$-calculus, and supporting artificial intelligence research.

1.  Functional programming. The hallmark of Lisp is that it was the first language to rely completely on recursion and functions. Lisp programmers of the 1960s and 1970s spoke of pure Lisp, which disallowed the use of programming structures such as `while` and `for` loops of imperative languages.
2.  The $\lambda$-calculus. As part of the Lisp development, McCarthy (1963) wrote "A Basis for a Mathematical Theory of Computation," in which he laid out the case for moving programming from the Turing machine model then in vogue to the $\lambda$-calculus version. His view would ultimately become the dominant view, and for that reason the $\lambda$-calculus basis is used in this text.
3.  Artificial intelligence. Until Lisp came along, there were no adequate languages for artificial intelligence research. Lisp became, and perhaps remains, the preeminent language in artificial intelligence research. Among its attributes, the inclusion of garbage collection made many artificial intelligence algorithms practical. Lisp 1.5 came out in 1962 from which sprang many different Lisp-like systems.

### 12.4.3  SNOBOL and Pattern Matching

SNOBOL was introduced in 1962 and it was based on the idea of pattern matching. At a time when built-in string manipulation operations were nonexistent, SNOBOL had a full repertoire of string operations, including full pattern-matching capability. It was dynamically typed and interpretive (as were the first Lisps). While not a direct descendent, Prolog shares SNOBOL's pattern-matching and searching focus.

### 12.4.4 APL and Matrices

Finally, a completely different language was developed: APL, short for "A Programming Language." APL was the invention of Kenneth Iverson and to say that it was unique is probably an understatement. Iverson had invented a notation for describing matrix computations, which he turned into a programming system in 1960. It was so unique that a special keyboard was required. Although completely unique, APL represents how far thinking ranged in the early 1960s as to what programming and programming languages were all about.

## 12.5 PROGRAMMING LANGUAGES FROM 1965 TO 1980

Although many programming languages were invented and many of the original languages underwent extensive development in this period, there were three events of notice: the development of structured programming, the implementation of Smalltalk, and the implementation of ML.

### 12.5.1 Structured Programming

Early programming languages and programming methodologies were dominated by a construct called the goto, something that is rarely seen in more contemporary languages. The construct does not even exist in Java, for example. Structured programming was based on a theorem in computability but it took the famous paper by Dijkstra (1968), "Go To Statement Considered Harmful," to popularize the notion. It is not worth recanting the details here, but suffice it to say that the elimination of gotos was seen as immensely important.

The upshot is that the languages of the day worked hard to remove, or at least minimize, the use of the goto construct. Perhaps the best-known of the "non-goto" languages was BLISS. BLISS is a system programming language developed at Carnegie Mellon University by W. A. Wulf, D. B. Russell, and A. N. Habermann around 1970. It was used extensively to develop systems until C. BLISS is a typeless block-structured language based on expressions rather than statements, and does not include a goto statement.

While not as radical as BLISS, Pascal made its appearance in 1970, along with C in 1971. These were noted earlier.

### 12.5.2 The Advent of Logical Programming

Prolog was introduced in 1970. Compared to either the imperative style or the functional style, Prolog is a distinct departure. Imperative and functional languages are based on λ-calculus; Prolog is based on logic. Hence, Prolog represents the archtypal logic programming language.

Prolog was initially designed to support artificial intelligence applications (recall Lisp). Prolog programs have two readings: a declarative one and an operational one. Prolog programmers place a heavy declarative emphasis on thinking of the logical relations between objects or entities relevant to a given problem, rather than on procedural steps necessary to solve it. The system decides the way to solve the problem, including the sequences of instructions that the computer must go through to solve it. It solves problems by searching a knowledge base (database) that would be greatly improved if several processors are made to search different parts of the database.

### 12.5.3  ML and Types

ML was originally developed as an experiment in type-checking. ML is a general-purpose functional programming language developed by Robin Milner and others in the late 1970s at the University of Edinburgh. ML stands for *metalanguage* as it was conceived to develop proof tactics in the LCF theorem prover. LCF stands for logic of computable functions and is important in its own right. The language of LCF was pplamda, for which ML was the metalanguage. ML is known for its use of the Hindley–Milner type inference algorithm, which can infer the types of most values without requiring the extensive annotation often criticized in languages such as Java.

### 12.5.4  Our Old Friend: Forth

Forth was introduced in 1968. Its history was given in Milestone I and it is mentioned here for completeness.

## 12.6  THE 1980s AND 1990s: THE OBJECT-ORIENTED PARADIGM

The major innovation in this timeframe was not an innovation as much as the popularization of the object-oriented paradigm. In these two decades we see the rise of C++, Objective C, Objective Pascal, Objective CAML (an offshoot of ML), Python (and its non-object base, Perl), and, of course, Java. The object paradigm derives its usefulness by naturally associating real-world objects with software models through the language. The popularity of the paradigm can be seen by reading the "help wanted" advertisements.

## 12.7  THE PARALLEL PROCESSING REVOLUTION

A true revolution, certainly in the United States, was the *HPCC*, the High Performance Computing and Communications Act, in 1991. This Act provided funds and momentum to develop parallel and distributed computing and what would become the Internet. Now things get interesting.

### 12.7.1 Architecture Nomenclature

One classification scheme of architectures is based on the number of instruction streams by the number of data streams, as shown in Figure 12.1. For exmaple, a simple von Neumann style chip has one instruction stream (as pointed to by the program word) and one data stream (it retrieves one piece of data at a time). Architectures meant to process mathematical matrix programs are often SIMD; the same instruction is simultaneously executed on many different data streams. SIMD architectures gain one order of magnitude in execution time; they effectively remove the innermost loop in matrix algorithms. SIMD systems can, and have, made good use of the classical programming languages, especially FORTRAN.

Things are more interesting when we consider multiple instruction streams. MISD architectures have been developed (Halaas et al. 2004), but by far the most familiar architectures are the Multiple Instruction, Multiple Data architectures. The Internet can be thought of as a massive MIMD system.

### 12.7.2 Languages for Non-SISD Architectures

One of the original languages suitable for MIMD work was proposed in 1974 by C. A. R. Hoare for operating systems. Hoare's proposal was for *message passing primitives* of `send` and `receive`. In order to control concurrency, `receive` was a blocking primitive—it caused a wait. Occam was the system used by the INMOS transputer (now defunct), but occam is still available.*

The development of operating systems has been a driver of programming language development, C being the canonical example. Language features, such as `monitors` (a type of object), was introduced by Hoare in 1974. Monitors encapsulated semaphores, but were not the complete langauge as was occam.

The programming language UNITY was developed by K. Mani Chandy and Jayadev Misra in their book *Parallel Program Design: A Foundation* (1988). Although it was considered a theoretical language, it focused on what, instead of where, when, or how. The peculiar thing about the language is that it has no flow control. The statements in the program

---

* One of the lessons of programming language history is that, once developed, a language will have adherents forever.

**Online Resources**

Online descriptions of the the following language systems are available

### General Purpose Imperative Languages

| | | | | |
|---|---|---|---|---|
| Ada | Algol | Basic | C | C++ |
| C# | COBOL | FORTRAN | Modula-2 | Pascal |
| Perl | | | | |

### Object-Oriented Languages

| | | | | |
|---|---|---|---|---|
| C++ | Eiffel | Java | Objective-C | Python |
| Ruby | Simula | Smalltalk | | |

### Functional Languages

| | | | | |
|---|---|---|---|---|
| Caml | Common Lisp | Haskell | ML | Ocaml |
| Scheme | | | | |

### Logic Programming Languages

Prolog

### Web Development Languages

| | | | | |
|---|---|---|---|---|
| ColdFusion | Delphi | JavaScript | PHP | PL/SQL |
| PowerBuilder | Visual Basic | VB.NET | | |

### Database Programming

| | | | |
|---|---|---|---|
| ABAP | Clipper | MUMPS | T-SQL |

### Miscellaneous

APL: Matrices and linear algebra
AWK: Pattern Matching
Tcl: Command scripting language
RPG: Venerable report writing language
SAS: Statistical processing
Logo: Educational language of the turtle
CLIPS: Rule-based programming
Maple: Computer Algebra System
Mathematica: Computer Algebra System
Rational Rose: System Development System

### Architecture Types

| | Single data | Multiple data |
|---|---|---|
| Single Instruction | SISD | SIMD |
| Multiple Instruction | MISD | MIMD |

**Figure 12.1  Architecture Classification by Instructions and Data**

run in a random order until none of the statements cause change if run. To discrete event simulation practitioners, UNITY was a natural view of parallel/distributed processing. About the same time, Ian Foster and Stephen Taylor (1990) developed Strand, a more functionally oriented system.

Eventually, it seems, every language paradigm was moved to the parallel and distributed world, demonstrating once again Chomsky's Hierarchy.

## 12.8 CRITIQUE OF THE PRESENT

When we consider a lineage diagram, such as `levenez.com/lang/`, it is clear that there are very few unique brands of programming languages. And although there seems to be no end of argument over the "best" programming language, there also appears no way to scientifically establish "best." In fact, one might claim that the focus on programming languages only that one sees in an undergraduate curriculum is very short sighted.

Focusing on "conventional" programming is also a dead end. For example, programming systems, such as Maple™, Mathematica®, and Matlab®, are based on inputting information in an algebraic format, yet these languages often do not make use of what we've learned about language because the user base does not want to change to wit, types.

As the computer has become more ubiquitous, the need for better interfaces with the systems becomes more important. A classical case is the destruction of the Iranian airliner by the *USS Vincennes* in 1988. At least part of the issue was the inability to adequately display information in order for the captain of the *Vincennes* to make an informed decision.

We should even expect programming itself to change. For example, the concept of visual programming has arisen several times over the course of the last 50 years. Can programming be eliminated via automatic programming?

# Appendix A

## REFLECTIONS BY ONE GOOD STUDENT

These example milestone reports are courtesy of Clint W. Smullen IV, who took a course based on this text in the fall of 2005. These notes are provided to give students an idea of what a report should look like.

The reports are Clint's; I have removed code and drawings that would trivialize the projects.

## A.1 MILESTONE I

### A.1.1 Metacognitive Comments

I first began to learn Gforth by translating extremely simple expressions that I know in C into Gforth, such as: $1 + 2$ into $1\ 2\ +$. I then proceeded to look at how to manipulate the stack using ., .s, DROP, ROT, etc. This was followed by learning how to use floating-point arithmetic and the large differences between the manipulation of integer and floating-point numbers. It was not until after class on Monday that I knew how to use variables. At this point, I was able to translate the first two cases of expressions into Gforth, except for the integer exponentiation. To learn how to create words and utilize conditional expressions, I wrote a new word in Gforth to compute integer exponentiation. This heavily used the stack manipulation functionality which I had previously studied. From studying these various aspects of Gforth and translating instructions from a language I know well to this new language, I have learned something of the pathway and issues that will arise in the process of writing a compiler.

## A.2 MILESTONES II AND III

[Authors note: In the semester that Clint participated, there was only one report covering both the scanner and parser.]

## A.2.1 Designing the Scanner

Starting with the basic regular expressions for each class of token, I developed finite state automata (FSAs) that process these inputs. The basic classes of input are integers, floating point, strings, and names. The name token type also incorporates booleans and files. These FSAs were extended into finite state machines.

## A.2.2 Implementing the Scanner

I programmed my scanner to read chunks from the input. And scan through the chunks looking for matches. First, it looks to see if the current input character is either a left or a right square brace, then it checks for an opening double-quote character. If it finds a double-quote character, then it will begin parsing for a string. Otherwise, the scanner extracts the longest possible input token from the input and tries to match it to an integer, float, or name. The C Standard Library functions `strtol` and `strtof` are used to check the input for being an integer or float. This ensures that the scanner "does as C does," when it comes to integers and floats. If the input token is not entirely an integer or a float, then it must be a name. At this point, I directly used the name token FSM to check this assumption. Once a set of data from the input stream is matched to the input, the data is stored into a token structure along with the matched type and returned back to the caller.

## A.2.3 Designing the Parser

I designed my parser around the simplified grammar that was discussed in class. This grammar moves the vast majority of the complexity inherent in the original grammar from the syntactic analysis that is occurring in milestone III, to the semantic analysis that will occur in later milestones. This simplified grammar serves only to rebuild the tree from the linearized version which is given as input to this compiler. The two necessary components center around the productions $\{S, T\} \rightarrow [\ T\ ]$ and $T \rightarrow T\ T$. The former of these two productions represents parsing the child of a node while the latter represents parsing a sibling. In a more compact form, the grammar can be made to show these two possible paths in parsing.

## A.2.4 Implementing the Parser

In programming my parser, I used the compact form of the grammar almost directly. The function that does the bulk of the parsing, called child-Parser, first checks for a left bracket token. If it finds one, then it recursively calls itself again until it sees a right bracket token. If it does not see a left bracket, then it directly saves the token that was read into the tree.

This function by itself implements only the productions starting at $T$, so, to complete the parser, I added another function, called `parser`, which calls `childParser` once and then checks that the node returned is a list and not an atom. These two functions form the same functionality described in class as the double-recursive algorithm, with the modification that one of the recursions is replaced by an iteration. Two functions are still necessary to obtain the correct input language.

## A.3   MILESTONE IV

### A.3.1   Designing the Symbol Table and Type Checker

First, it was necessary to decide upon what data will be stored in a symbol. For function symbols, the name of the function in both SOL and Gforth are necessary, as is the return type and the types of all of the arguments. Variable symbols need the name, the type, and a value for static variables such as true, false, stdin, etc. The symbol table structure is an array of pointers to symbols. There will be two arrays for each type of symbol: variables and functions. Though this will require more duplicated code, it simplifies the design, because there is no ambiguity about what the type of a symbol is.

Starting with the premise of a hash table, I looked at different possibilities and decided on using open-addressing with quadratic probing. By keeping the hash table less than half full and using prime table sizes, it is always possible to find a hashing location with quadratic probing. Following onto this, it was necessary to design hash functions to hash the symbol structures as well as a node from the parse tree and have the hash values match. [Author's note: this is completely counter to the requirements!]

I decided to use the string hashing algorithm described in Mark Allen Weiss's *Data Structures and Algorithm Analysis in JAVA*. This algorithm is similar to the algorithm used in the Java libraries for hashing. Although this algorithm alone is sufficient for computing the hash values for variable names, it is not quite enough to hash the complete input specification for a function. To incorporate the effect of arguments, I decided to extend the use of the hashing algorithm above by additionally hashing the types, in numeric form, at the end of the name string. When running a parse tree through the type checker, all of a function's arguments will have been typed before looking it up in the symbol table, so it is universally acceptable to use this method. When hashing a symbol, all of the types are immediately available and, when hashing a parse tree node, the types of each argument will have already been determined.

The type checker is a bottom-up algorithm, as discussed in class, which checks all of the children of a node before performing type checking on the node itself. Upon encountering a token node in the parse tree, the type checker will use its type directly for non-name tokens or attempt to look its type up using the name given. To type check a list node with the contents $a_0 \, a_1 \ldots a_n$. The types of the children $a_1 \ldots a_n$ will first be checked recursively. An error is thrown if $a_0$ is not a name token, otherwise, the symbol table is searched for a symbol matching the name $a_0$ with matching the count and type of $a_1 \ldots a_n$. If no symbol is found, then an error is thrown, otherwise a pointer to the symbol is saved into the parse tree. This way, both the name of the Gforth library word and the return type are immediately available without any need to add or alter the data in the tree.

### A.3.2 Implementing the Symbol Table and Type Checker

I implemented the structure of the symbol table exactly as designed along with hashing routines, hash table resize and rehashing routines, as well as routines to print out the contents of the symbol table. The hashing routine was implemented as designed, with the modification that when combining on the effect of function types, the previous iteration's value is multiplied by 67 instead of 31. This gave improved distribution and reduced the overall number of collisions.

At least for this milestone, the library function prototypes are being manually added to the symbol table when the parser is initialized. This requires a great deal of repetitive code into which typographical errors can easily creep. The ability to print out the symbol tables helped greatly with debugging these errors.

The parser's print tree routine was extended so that it would print out the types applied by the type checker, as well as printing out the Gforth library word name instead of the input name token for the head of a list node. However, all of the existing tree printing functionality was preserved so that it is possible to print out a tree that has yet to be type checked or has been only partially type checked. This, as with the symbol table printing routines, was extremely helpful in debugging the type checker and ensuring correct operation.

### A.3.3 Testing Concerns

By using the print routines for both the trees and the symbol tables I was able to make sure that the data stored in various data structures was correct. I have not, as of yet, uncovered any potential pitfalls for future milestones except for the present implementation of the library prototypes. Since I have already begun considering the next milestone, I am already working on the design of a system to read the prototypes from a file at startup, eliminating the huge source of potential error in the code that manually inserts entries into the symbol tables for library functions. The keyword constants, such as true and false, will continue to be manually inserted by the code, since they are actually part of the language more than part of the Gforth library.

## A.4 MILESTONE V

Performing the actual translation to Gforth, at this step, was quite easy. As discussed in class, for the case of constants-only SOL input, all that is necessary is outputting into postfix. It was necessary for me to take special care when handling strings, however, as Forth only guarantees that a single immediate string, as specified by sör s stand and", will be held in memory at a time. To solve this issue, I used string utility functions that I had already written for use in manipulating strings in my library, strcpy and createString. To output a string constant, the translator first outputs a createString word prefixed with the correct length, then the constant string is output, using the C-style escape format, and then a strcpy word is output, copying the temporary string into heap storage. Of course, a great deal more work than this will be necessary to implement flow control structure and scope-producing statements that go along with functions and variables.

### A.4.1 Implementing a Library in Gforth

Implementing the arithmetic and comparison operators in Gforth is extremely easy in general, since it is of course necessary for Gforth itself to support the operations. Doing the type conversion and argument reordering is uniform across each input type format of the operator. Because Gforth does not have an integer exponentiation or floating-point modulus operator, it was necessary to implement those myself. The integer exponentiation word was already implemented for the first milestone, so I used that directly, but for the floating-point modulus it was necessary to implement it anew. The string operations took thought and a great deal of planning the stack effect. Initially I created the utility functions createString and strcpy. Using these I created the string concatenation operator + which first creates a new empty string with the length of the sum of the two input strings' lengths. It next uses strcpy to copy the first string into the first part of the new string and then uses strcpy to copy the second string to the space following the first string. The strings are stored in the preferred Gforth format of a (c-addr u) pair, so no null termination is necessary. To implement the string comparison operations, I implemented the C library's strcmp directly, so it returns -1 if string2 is "greater" than string1, 1 if string2 is "less" than string1, and 0 if they are equal. Using this function, implementing each of the comparison types was trivial. The only two remaining string functions are insert and charat. Both of them work by addressing a specific character in a string and returning the string back. The charat word simply reads out the contents of the memory location while insert replaces the character at the location with the given character. Since we decided in class to leave off the file I/O words, I did not implement them.

## A.5 REFLECTION

Implementing the tree to Gforth translation did not take any particular effort on my part, but it was necessary to take care when designing the stack effect for the more complicated library words. In particular the string functions, since it required pointer arithmetic without the use of registers. The only safe way I discerned it possible to design these functions was to determine the "target" stack state that I needed to perform an operation (like strcpy or +), and then try to determine the least number of operations necessary to reach that state. This worked out very well for me, and made serious debugging unnecessary. In going along with this design methodology, I included step-by-step stack effect descriptions at key points, making it possible for me to understand what is occurring at a later stage. Commenting like that helps me later on once I have forgotten why I coded something a certain way.

## A.6 MILESTONE VI

In approaching the front-end side of user-defined variables, I used the approach of separating each scope into its own symbol table. For my data structures, this was easiest, since it did not require any deletions. In determining the type for a variable, the user can either explicitly assign it using the [: <name> <type>] format or the type checker will automatically determine it from the type of the initial value expression. Either way, a new entry is placed into the appropriate symbol table and one of two stack offset counters is adjusted to reflect the new local variable. For this milestone, each scope was treated as a stack frame, so the only method that needs to perform the stack simulation is the instance of the type checker function which is type checking the let block. Begin-end blocks also create their own "scope," but it is not possible to declare any variables in it, because it is then necessary to use a let construct to actually do the declaring. For this reason, only let constructs do any scope or stack frame creation and deletion.

## A.6.1    Stack Frame and Gforth

To develop an operational model for dealing with stack frames in Gforth, I first sat down with Gforth and the various stack pointer words such as sp@, sp!, fp@, etc. After studying their operation, I decided to store both the argument and floating point stack points together in a double word variable called _#CURFRAMEPTR. The code I wanted to use to produce this value was sp@ fp@, so, because the Gforth stack grows down, the floating point frame pointer is stored in the lower half of _#CURFRAMEPTR while the argument stack frame pointer is stored in the upper half. Working with this design in mind, I developed routines to make stack frames, delete them without returning a value, and deleting them returning one of the SOL acceptable types as a return value.

In creating a frame, the current value of _#CURFRAMEPTR is pushed onto the stack, taking two cells. After this, the new stack frame pointers are stored into _#CURFRAMEPTR. Instead of "allocating" space on the stack for the total space needed for all variables, I simply had the initial value code for each local variable produce its value and place it on the appropriate stack. In returning from a stack frame, the previous stack frame pointers are pushed onto the top of the stack and then written directly to the stack pointers using sp! and fp!. Though an extremely unlikely situation to occur in a functional language, it prevents the possibility of accidentally having too many elements left on the stacks, causing the machine to go into an unknown state upon leaving a stack frame. If a value needs to be returned, it is copied into a global holding variable, the stack frame is destroyed, and then it is pushed back onto the appropriate stack. Though this solution may not be as refined as it possibly could be, it does the job in less running time and coding time.

## A.6.2    Variable Assignment

The other new issue in this milestone is variable assignment. To simplify the process on the type checker side, a function prototype was added to the library for each assignment operator. This function looks like a normal function to the type checker and thus requires no changes whatsoever to work with. The translator, however, must recognize that the destination needs to be passed by reference instead of by value. Correspondingly, the actual Gforth words that perform the assignment expect the first argument to be a pointer to the destination rather than the variable's value. This does put the entire job of handling variable assignment onto the shoulders of the translator, but it simplifies working with the code because there is only one component to worry about.

### A.6.3   Reflection

Working with scopes inside of my compiler was not difficult, but it took a great deal of time and effort to work through and debug the stack frames in Gforth. In the x86 architecture, all arguments are passed on the stack, as with Gforth, but there is only one stack. The most complicated part, for me, of working with stack frames in Gforth is trying to keep both stacks in the appropriate state. Since stack frame pointers also point to stack frame pointers, I had to be careful about how many pointers I dereferenced. Going one stack frame too far, especially when testing in a shallow set of stack frames, easily results in going below the bottom of the stack and getting hate mail from Gforth. Once I worked out each set of functions, I then just had to create bizarre nested functional code to try to break my compiler. I made sure to test cascaded lets, the situation where a child let hides the existence of a local variable in a higher scope, and also assignments. It is truly interesting to have written a compiler that does even this much.

### ·   A.7   MILESTONE VII

This was quite a challenge to get working correctly. I started out, as with local variables, working with Gforth, figuring out how I should implement it, and then writing words to encapsulate that implementation. I then set about adapting the type checker to correctly type check functions. Previously, for milestone VI, I had allowed the user to implicitly type variables. This became a problem in this milestone, because it is difficult to tell if the user is trying to declare a new variable or call a function with a single argument. Because of this, I mandated the explicit typing syntax for both variables and functions. This simplified the code, because there were far fewer possible variations that had to be dealt with. Additionally, I allowed the variable and function declarations and expressions inside a local scope let to be ordered any way the user desires. This too reduced the code required to perform the type checking and made it easier to integrate function definitions into my compiler.

In translating functions, I discovered for myself that Gforth does not allow for variable or colon definition of new words inside a colon definition, so it was necessary for me to go back and "float" prototypes of locally defined functions into the outermost parent function. With this in hand, I had the translator output a Gforth "defer" for each function, as well as outputting the function frame pointer variable declaration. This, combined with name mangling, provides an effective method for dealing with locally defined functions in SOL.

In Gforth, I created a library word for padding the current stack frame, so as to make space for the local variables, and altered the destroyFrame words so that they took the extra offset needed to clear off the arguments on the two stacks. The procedure is then to first create the stack frame, which saves and updates the previous global stack frame pointer, then to save and update the function frame pointer, which is used by other functions to get at local variables not belonging to themselves. The main issue with my solution is that it took a fair amount of work to determine what to start the offsets at and how much to offset by to get at the arguments.

### A.7.1 Reflection on This Milestone

This milestone took a great deal of testing and debugging to complete. Most of the test cases that I created uncovered a problem, which then had to be fixed. Misalignment of offsets and the improper manipulation of variables were the main issues. My schema for where local variables are and how they are accessed changed significantly from milestone VI, but some of my code was not properly updated to reflect that change. Some way of associating code with its design element would have helped me to resolve the issues sooner. The largest overall issue for me was the fact that there were three languages involved when doing any testing, each with a complete set of semantics. On previous milestones, the input complexity was very limited, also limiting the output complexity which made it easier to figure out what was wrong at some point. Here, the issue could be, if the compiler is throwing an error, that either the input file or the compiler has an error. That kind of error is somewhat easier to deal with by looking at a very familiar programming language, C, and analyzing its operation to figure out what is wrong. Because of this, it has been easy for me to fix problems without directly encountering them, merely anticipating them when I am going through the code looking for a separate error. If the output program does not work, then it could be any of the three, and it takes me a great deal of time to determine the exact problem. Gforth does not give line numbers for errors inside compiled words, so it is necessary to insert print statements to determine where it is exactly. Then it is necessary to examine the stack before and after certain events, to make sure that the correct number and type of values are being generated, and finally I frequently have to look at the semantics for my library words to make sure that what it should be doing corresponds to what I expect it to be doing.

## A.8 REFLECTION ON THIS COURSE

I find compilers to be a very interesting subject. This course was very challenging and time intensive. The only reason I completed through the final milestone was that I always worked to stay on top of all of the milestones, which did become progressively harder throughout the semester. It seems that this semester has, for everyone I know both in and out of this course, been the busiest yet, making it extremely difficult to keep on top of the workload for this course especially. It would have been extremely helpful if we had gotten an earlier start on the earlier milestones and finished them sooner, giving more time for the later milestones, which truly require it. It might also have been useful to have gone through all the components of the compiler sooner and worked on designing the complete solution, rather than piecemeal. Doing as it was done did, however, require a great deal of thought at each point to figure out how to go from where I was to where I needed to be. The required course book was not very useful through this process. I found other books at the library that actually described the components required to implement a compiler which were more helpful in designing my milestones. The project book would be more useful if it had explicit examples of the SOL code in use and if the milestone descriptions actually described what was necessary for completion of the milestones.

# REFERENCES

Aho, Alfred V. and Jeffrey D. Ullman. 1972. *Theory of Parsing, Translation and Compiling*. Englewood Cliffs, NJ: Prentice Hall.

Aho, Alfred V., Ravi Sethi, and Jeffrey D. Ullman. 1986. *Compilers: Principles, Techniques, and Tools*. Reading, MA: Addison-Wesley.

American National Standards Institute (ANSI)/Institute of Electric and Electronics Engineers (IEEE). 1993. X3.215-199x. Draft Proposed American National Standard for Information—Programming Languages—Forth. X3J14 dpANS-6—June 30.

Anderson, L. W. and D. R. Krathwohl. 2001. *A Taxonomy for Learning, Teaching, and Assessment: A Revision of Bloom's Taxonomy of Educational Objectives*. New York: Longman.

Barendregt, H. P. 1984. *The Lambda Calculus: Its Syntax and Semantics*. Amsterdam: North Holland.

Bloom, B. S. 1956. *Taxonomy of Educational Objectives, Handbook I: The Cognitive Domain*. New York: David McKay.

Boehm, Corrado and Guiseppe Jacopini. 1966. "Flow Diagrams, Turing Machines and Languages with Only Two Formation Rules." *Communications of the ACM* 9 (May): 366–71.

Bransford, John D., Ann L. Brown, and Rodney R. Cocking, eds. 2000. *How People Learn: Brain, Mind, Experience, and School*. Washington, D.C.: National Academies Press.

Brodie, L. 1981. *Starting FORTH*. Englewood Cliffs, NJ: Prentice Hall.

———. 1984. *Thinking FORTH*. Englewood Cliffs, NJ: Prentice Hall.

Brooks, F. P. 1975. *The Mythical Man Month*. Reading, MA: Addison-Wesley.

Chandy, K. Mani and Jayadev Misra. 1988. *Parallel Program Design: A Foundation*. Reading, MA: Addison-Wesley.

Chomsky, Noam. 1956. "Three Models for the Description of Language." *IRE Transactions on Information Theory* 2: 113–124.

———. 1959. "On Certain Formal Properties of Grammar." *Information and Control* 1: 91–112.

Church, Alonzo. 1936. "A Note on the Entscheidungsproblem." *Journal of Symbolic Logic* 1: 40–41.

———. 1941. *The Calculi of Lambda-Conversion*. Annals of Mathematics Studies. Princeton, NJ: Princeton University Press.

Crimmins, Mark. 1998. "Language, Philosophy of." In *Routledge Encyclopedia of Philosophy*, ed. E. Craig. London: Routledge (retrieved February 2005, from http://www.rep.routledge.com/article/U017).

Dijkstra, Edsger W. 1968. "Go To Statement Considered Harmful." *Communications of the ACM* 11(3): 147–48.

Fleming, Quentin W. and Joel M. Koppelman. 2000. *Earned Value Project Management*, 2nd edition. Newtown Square, PA: Project Management Institute.

Foster, Ian and Stephen Taylor. 1990. *Strand: New Concepts in Parallel Programming*. Upper Saddle River, NJ: Prentice Hall.

Halaas, A., B. Svingen, M. Nedland, P. Saetrom, O. Snove, Jr., and O. R. Birkeland. 2004. "A Recursive MISD Architecture for Pattern Matching." *IEEE Transactions on Very Large Scale Integration (VLSI) Systems* 12(7): 727–34.

Hankin, C. 2004. *Lambda Calculi: A Guide for Computer Scientists*. London: King's College Publications.

Heuring, Vincent P. and Harry F. Jordan. 2003. *Computer Architecture and Organization: An Integrated Approach*. New York: John Wiley & Sons.

Hoare, C. A. R. 1969. "An Axiomatic Basis for Computer Programming." *Communications of the ACM* 12(10): 576–80.

———. 1974. "Monitors: An Operating System Structuring Concept." *Communications of the ACM* 17(10): 549–57.

Humphrey, Watts S. 1996. *Introduction to the Personal Software Process*. Reading, MA: Addison-Wesley.

Jones, Richard and Rafael Lins. 1996. *Garbage Collection: Algorithms for Automatic Dynamic Memory Management*. New York: John Wiley & Sons.

Kleene, S. C. 1936. "Lambda-Definability and Recursiveness." *Duke Mathematical Journal* 2: 340–53.

Louden, Kenneth C. 2003. *Programming Languages: Principles and Practice*, 3rd edition. Belmont, CA: Thompson, Brooks/Cole.

McCarthy, John. 1963. "A Basis for a Mathematical Theory of Computation." Pp. 33–70 in *Computer Programming and Formal Systems*, ed. P. Braffort and D. Hirschberg. Amsterdam: North Holland.

Moore, C. H. 1980. "The Evolution of FORTH—an Unusual Language." *Byte* (August).

National Research Council. 2000. *How People Learn: Brain, Mind, Experience, and School*, expanded edition. Washington, D.C.: National Academies Press.

Paulson, Lawrence C. 1996. *ML for the Working Programmer*. London: Cambridge University Press.

Plotkin, Gordon, D. 1981. *A Structural Approach to Operational Semantics*. Technical report DAIMI FN-19. Computer Science Department, Aarhus University, Aarhus, Denmark.

Pólya, George. 1957. *How to Solve It*, 2nd edition. Princeton, NJ: Princeton University Press.

Pratt, Terrence W. and Marvin V. Zelkowitz. 1995. *Programming Languages: Design and Implementation*, 4th edition. Upper Saddle River, NJ: Prentice Hall.

Russell, Bertrand. 1908. "Mathematical Logic as Based on the Theory of Types." *American Journal of Mathematics* 30: 222–62 (reprinted 1967, in Jean van Heijenoort, *From Frege to Gödel*, pp. 152–82. Cambridge, MA: Harvard University Press).

Simon, H. A. 1969. *The Sciences of the Artificial*. Cambridge, MA: MIT Press.

Turing, A. 1936. "On Computable Numbers, with an Application to the Entscheidungsproblem." *Proceedings of the London Mathematical Society, Ser. 2* 42: 230–65.

Weiss, Mark Alan, Data Structures and Algorithms in JAVA. Additions-Wesley 2003.

# INDEX